Bernard Stonehouse
Cambridge
1992

AF841438

ANTARCTICA

KEITH SUTER

ANTARCTICA

Private Property or Public Heritage?

KEITH SUTER

Pluto Press Australia

London and New Jersey

Antarctica: Private Property or Public Heritage? was first published in 1991 by:

In the United Kingdom and the United States of America
Zed Books Ltd, 57 Caledonian Road, London N1 9BU, and 171 First Avenue, Atlantic Highlands, New Jersey 07716

In Australia and New Zealand
Pluto Press Australia, P.O. Box 199, Leichhardt, NSW 2040

Copyright © Keith Suter, 1991

Map supplied by the Scott Polar Research Institute
Cover design by Sophie Buchet
Cover picture shows Antarctica Expedition 2nd leg. Antarctica landscape with MV Gondwana in distance. Cape Evans. 30 January 1990. Copyright Greenpeace/Midgley.
Typeset in North Wales by Derek Doyle & Associates, Mold, Clwyd.
Printed and bound in Great Britain by Biddles of Guildford and King's Lynn.

British Library Cataloguing in Publication Data
Suter, Keith
Antarctica : private property or public heritage?
1. Antarctica
I. Title
333.782160989

ISBN 0-86232-846-2
ISBN 0-86232-847-0 pbk.

Library of Congress Cataloging-in-Publication Data
Suter, Keith
Antarctica : private property or public heritage? Keith D. Suter.
p. cm.
Includes bibliographical references.
ISBN 0 86232 846 2. ISBN 0-86232-847-0 (pbk.)
1. Antarctic regions—International status. 2. Public domain – Antarctic regions. 3. Mines and mineral resources—Antarctic regions. I. Title.
JX4084.A5896 1989
333.1′0998′9—dc20 89-35867
CIP

Australian ISBN 0 949138 65 7

Contents

	Abbreviations	viii
	Preface	ix
1.	**The Unique Continent**	1
	The importance of Antarctica	1
	The last discovery	3
	Antarctica and the Second Environmental Wave	5
	A window of opportunity	7
2.	**The 1959 Antarctic Treaty: No Cold War at the South Pole**	9
	Good news is not news	9
	What is International Law?	9
	Where does International Law come from?	10
	The acquisition of territory	13
	National claims to Antarctica	15
	International Geophysical Year	17
	The 1959 Treaty	19
	An assessment of the Treaty	20
	Limitations of the Treaty	22
	A visionary treaty	24
	Annex to Chapter 2: Antarctic Treaty Nations	27
3.	**Antarctic Living Resources**	28
	Whales	28
	Antarctic seals	32
	Agreed measures	34
	Antarctic living resources	34
	Conference on the Conservation of Antarctic Marine Living Resources	37
	CCAMLR – an assessment	39
	CCAMLR in action	41

4.	**Antarctic Minerals**	46
	'Don't sign – don't mine!'	46
	An Antarctic El Dorado?	47
	Antarctic minerals	48
	Creating the Minerals Convention	52
	Antarctic Minerals Convention	55
	Initial reactions to the new convention	57
	Australia's refusal to sign	59
	France opposes the treaty	63
	New Zealand's new policy	65
	The UK and US	66
	The window of opportunity	67
5.	**Antarctica on the International Agenda**	69
	Outside the Antarctic club	69
	The Law of the Sea Conference	69
	The 1975 Sri Lankan initiative	72
	1981: India's Operation Gangotri	73
	1982: Malaysia's UN speech	75
	Why Malaysia?	76
	UN General Assembly resolutions	78
	The UN's glacier	86
	White Australia	87
	A new sense of urgency	92
6.	**Coming in from the Cold**	95
	Antarctica's new era	95
	The South African connection	96
	The French connection	100
	Antarctic waste disposal and oil spills	108
7.	**The Greening of Antarctica**	112
	International commissions	112
	World conservation strategy	113
	Our common future	115
	The gap in the ozone layer	118
	The greenhouse effect	121
8.	**Antarctica and the Fifth World**	127
	The Fifth World	127
	The neglected subject	128
	The work of NGOs at the international level	131
	The Fifth World and Antarctica	133
	The window of opportunity	137

9. No Longer the neglected Continent 144
Here to stay 144
Thinking about proposals for Antarctica's future 145
Applying the six questions 146

10. Can We Build Upon the Antarctic Treaty System? 149
A durable treaty system? 149
A tourist treaty 150
A warning to tourists 152
Harvesting the ice 155
An Environmental Protection Agency 157

11. The Common Heritage of Humankind 159
Words are bullets 159
The tragedy of the commons 159
The common heritage of humankind 160
Lawyers all at sea 161
Two contrasting concepts 162
The status of Antarctica 163
Limitations of the common heritage concept 165

12. The Public Heritage of Humankind 169
No easy answers 169
The concept of the Public Heritage 170
Benefits of the Public Heritage to the Consultative Parties 176
Benefits of the Public Heritage to the planet 178
Campaigning for Antarctica – a job for life 179

Further Reading 182

Appendixes
I: The Antarctic Treaty 183
II: Convention on the Conservation of Antarctic Marine Living Resources 189
III: Extract from the Final Act of the Conference on the Conservation of Antarctic Marine Living Resources 203

Index 205

Abbreviations/Acronyms

AAT	Australian Antarctic Territory
ASEAN	Association of South East Asian Nations
ASOC	Antarctic and Southern Ocean Coalition
ATCP	Antarctic Treaty Consultative Parties
CCAMLR	Convention on the Conservation of Antarctic Mineral Resources
CFCs	chlorofluorocarbons
CND	Campaign for Nuclear Disarmament (UK)
CRAMRA	Convention on the Regulation of Antarctic Mineral Resources
DAPG	Dolphin Action and Protection Group (South Africa)
DEA	Department of Environment Affairs (South Africa)
ECOSOC	Economic and Social Council
EEZ	Exclusive Economic Zone
EIA	Environmental Impact Assessment
ICRC	International Committee of the Red Cross
IGY	International Geophysical Year
IJC	International Court of Justice
IUCN	International Union for the Conservation of Nature and Natural Resources
IWC	International Whaling Commission
NIEO	New International Economic Order
OPEC	Organization of Petroleum Exporting Countries
SCAR	Scientific Committee on Antarctic Research of the International Council of Scientific Unions
UNEP	United Nations Environment Programme
WCED	World Commission on Environment and Development
WILPF	Women's International League for Peace and Freedom

Preface

A society is judged not only by what it achieves but also by what it aspires to achieve. I believe that the current global society will be judged partly by what it did to protect Antarctica, for the 1990s are particularly crucial years for Antarctica. After so many decades – if not centuries – of being ignored by politicians, Antarctica is now on the international political agenda. Antarctica has come in from the cold.

I have followed Antarctica affairs for over a decade. I first became intrigued by the Antarctic Treaty as an example of a treaty operating well – and as a consequence, attracting little attention from the media or scholars. Good news is not news. But I also became aware that the political foundation of the Antarctic Treaty was not a firm one and that the Treaty would eventually run into problems. My warning in *World Law and the Last Wilderness*, first published a decade ago, was seen at the time as too alarmist. Reading the two editions in the light of current events, I seem now to have erred on the side of caution. The Antarctic problem is not going to disappear from the international agenda.

This book has three main purposes. The first is to revamp entirely my earlier book.

For too long, Antarctica has been the small preserve of scientists, lawyers and diplomats. It could do with a little 'glasnost'. Consequently, this book's second purpose is to provide an introduction to Antarctica's law and politics with a view to assisting and possibly even generating some public debate.

The third purpose is to convey some of the flavour and importance of international law. International law is often a cinderella subject at university (and invisible at school). Law students are not necessarily obliged to study it and it is often overlooked in courses on international politics. But the international legal order is confronted with new and unprecedented dangers. Citizens have a vested interest in learning more about international law. But many international law books are esoteric, bland and uninviting. This book is an example, I hope, of how international law can be conveyed as an interesting subject.

I would like to thank Mrs Myrilla Radnedge and Mrs Lynda Langton-Smith of Trinity Peace Research Institute for their typing. I am also grateful to Stephanie Dowrick of Pluto Press and Robert Molteno of Zed Books for their comments on the draft.

Keith Suter
Perth, 1990

Antarctica: seas, islands, lands and ranges

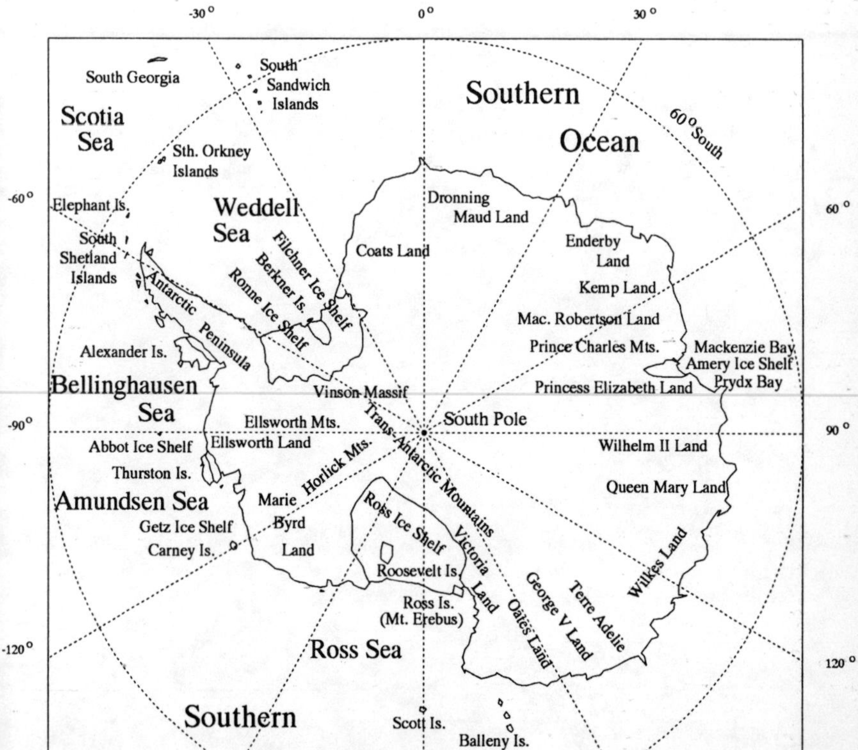

1. The Unique Continent

The importance of Antarctica

The world has only two frontiers left: the sea-bed and Antarctica. Like the sea, Antarctica is now being made vulnerable to exploitation because of advances in technology and the exhaustion of the more easily accessible resources elsewhere on earth.

Antarctica is currently on the international agenda. It is controlled by 25 Consultative Parties with one feature in common: a desire to maintain their control over the Antarctic and not to admit wider involvement in it by the international community. This monopoly is no longer acceptable to some Third World countries who are using the annual sessions of the UN General Assembly as a way of trying to open up control of the Antarctic to the international community. Meanwhile environmental non-governmental organizations are also becoming more active in Antarctic affairs.

The coming decade will see whether the continent will remain under the control of a comparatively small number of nations which have a unique legal arrangement for governing it. Alternatively, that unique arrangement may be replaced by some other arrangement under which the continent is governed on behalf of all humankind.

Antarctica demonstrates a particularly complex mixture of political and legal problems. Antarctica contains many instances of the main themes of international politics: colonialism; transnational corporations and technologically developed nations benefiting at the expense of developing nations; exploitation versus conservation; and North versus South. All of these issues impact on nations in different ways. For example, the USSR wants to develop its fishing industry but is less enthusiastic about an Antarctic oil and gas industry, both because its own energy technology is not as advanced as, say, the US one, and because as a supplier of oil and gas it does not mind if potential competition stays in the ground. The US's technological position is the direct reverse – but it still has a dilemma over exploitation: it wants the oil for itself and its transnational corporations while at the same time having to be aware of the votes to be gained at home through being seen as a conservationist nation in Antarctica. The superpowers also have a joint dilemma: how can they justify their arms race when they co-operate so well in the Antarctic? Chile is another example. It claims to control some of the continent and so would like to benefit from mineral development. But its fisheries are acutely dependent upon the Antarctic ecosystem and a massive oil spill could cause great damage to that industry (oil biodegrades much more slowly in cold latitudes than in the hotter latitudes so the oil slick would last longer here than elsewhere).

Meanwhile, Antarctica is the world's last wilderness. There are various moral arguments that can be advanced to support the view that the last wilderness should be preserved *as the last wilderness*. At the very least, future generations should be given the same benefits from environment as the present generation. Additionally, there is a new view developing that the environment has a right to exist irrespective of its utility to humans. Just how this claim could be pressed at law remains unclear. But this is certainly a growth area in philosophy and no doubt in the coming years we will hear far more of what must now be called 'environmental liberation' (following hard on the heels of and often concomitant with 'animal liberation').

Antarctica has a unique role in scientific research. It is a control area against which we can measure environmental deterioration in other parts of the globe. Indeed, its scientific uses continue to grow each year. In recent years, for example, it has played a vital role in our knowledge of meteorites. Most of the meteorites which are now being investigated have been located in the ice at the Antarctic. In addition, isotopic dating techniques indicate that, on average, the Antarctic specimens fell to earth much earlier than those collected on other continents. The great majority of meteorites found outside Antarctica fell within the last 200 years, although a few have been discovered that arrived tens of thousands of years ago. Yet most of the Antarctic stones have apparently been there for about 1,000–750,000 years. There has been ice at the South Pole for at least 15 million years and it is hoped that eventually meteorites which fell to earth more than a million years ago will be located deep in the ice.

The Antarctic, incidentally, is of use not only to scientists concerned with the earth but also to astronomers. In order to observe activity on the sun's surface without interruption, it is necessary to have a 24-hour constant observation of the sun. Other than going into orbit around the earth, the only alternative is to be based at either the North or South Poles during their respective summers. This was done in a major way during International Geophysical Year and it has since been continued because of its initial success.

Another example of scientific research concerns the origin of life on earth. Fossil bones of a plant-eating dinosaur found in Antarctica in early 1989 show there was a milder climate there 70 million years ago. 'We discovered jawbones, much of the vertebral column and the forelimbs, which identify the animal as a bird-hipped dinosaur similar to hypsilophodon, which has been found in Australia, North America, Europe and possibly North Africa,' said Mr Michael Thomson, a scientist who led an expedition to the area.[1] 'There could not have been much ice or snow in the region in those days because the dinosaur was a land animal and it wouldn't have liked frosty nights.' The dinosaur was about three metres long and walked on its hind legs. The expedition also found shells, fossilized leaves, tree trunks and ferns, indicating there was a food source for the dinosaur. There were active volcanoes nearby in those times, and the area would have been something like Indonesia is today.

Antarctica plays an important role in the regulation of the earth's weather pattern. Indeed, the Antarctic ice sheet would make a good study for a horror film. There is a continuing debate among scientists regarding the 'greenhouse effect' or whether the earth is warming. Larger than all of Europe, the Antarctic ice sheet would, if it slid, cover a large part of the southern Atlantic and Pacific Oceans. If it melted, the sea level would rise by at least 1.5 metres around the world, thereby flooding coastal areas.

Finally, there is a more immediate threat of what exploitation could mean for the Antarctic. There is the potential over-exploitation of krill, which would hurt the whale populations, and there are dangers in trying to extract the non-living resources. But the cost of extraction would be extremely high so there is no immediate possibility of any commercial exploitation taking place. The world's economy would need to be in a very different situation from its present state to justify the very expensive products that would be mined at the Antarctic, although much the same has been said about North Sea oil. When exploitation began there, tentatively, more than 20 years ago, the technology involved was then the most advanced of deep-sea mining technology and many economists claimed that North Sea oil would never be profitable because of the then declining cost of Middle East oil. That situation changed abruptly in 1973 following the increase in OPEC prices. Since that time the UK has benefited considerably from its exploitation of the sea-bed. Consequently, although there is no immediate risk of exploitation of the non-living resources in the Antarctic, the risk cannot be ruled out.

The last discovery

The Greeks knew that Antarctica existed. Thousands of years ago, having noted the existence of the Arctic, they assumed that there must be a similar countervailing continent to the South. That continent, however, remained distant from human contact until about 200 years ago. It has been only in this century that major explorations have taken place, some of them involving loss of life. The Greeks had the right answer from the wrong reasoning. The Arctic is simply a sheet of ice, under which it is possible for submarines to travel. The Antarctic is rock covered with a mantle of ice.

Antarctica is unique among the world's continents in that it is almost entirely covered by an ice sheet. It is estimated that there are about 26 million cubic kilometres of ice in Antarctica, enough to cover the whole of Australia with a mantle nearly four kilometres thick. The dome-shaped ice sheet has been formed by the accumulation of snow over millions of years. At its highest point, the ice is over four kilometres above sea level, and beneath it is a hidden landscape of mountains, valleys and plains. This ice represents about 65 per cent of the world's total supply of fresh water.

The ice is forcing Antarctica's land mass down into the surface of the Earth. Its weight distorts the planet's shape so that earth is not a flattened sphere but a pear shape. Every year the Antarctic ice sheet sheds icebergs; these are usually as large as ocean liners and sometimes as small as a family car. The biggest iceberg seen was spotted in 1956. At 335 kilometres long and 97 kilometres wide, it was the same size as Belgium; it took years to melt.

Antarctica was the last continent to receive human contact. It was not until the 1820s that the Antarctic mainland was first sighted. For over a century, the handful of people interested in Antarctic affairs were explorers on land and whalers active in the extremely fertile Southern Ocean. Seven nations claimed various parts of the continent: Australia, France, Chile, Argentina, New Zealand, the UK and Norway. Three of the claims (those of Chile, Argentina and the UK) actually overlap. Overlapping territorial claims are a recipe for conflict elsewhere on the globe. Antarctica has, however, been free of armed conflict. Until a few years ago, technology did not exist to create and maintain garrisons on the continent, and there was certainly no permanent population.

Even today, Antarctica is not a 'settled' land mass in the customary sense. There is no permanent population – although Chile and Argentina are encouraging some of their citizens to reside there for periods amounting to a few years. The 'garrisons' are staffed largely by scientists. The military are deployed there to provide logistic support only. Antarctica is the world's first demilitarized continent; it is also the first continent in which scientific exploration has overshadowed military deployments.

Antarctica and the Second Environmental Wave

What is urgent will always take precedence over what is important. In all human affairs there seem to be so many pressing matters that the 'urgent' ones get priority and so 'important' ones are put to one side 'temporarily' – but are rarely addressed because other 'urgent' matters come along.

Antarctica, and environmental matters generally, have usually been 'important' but have often been overshadowed by 'urgent' matters closer to home such as unemployment, inflation and the energy crisis. The late 1980s, however, have seen the rise of both Antarctica in particular and environmental matters generally as both 'important' and 'urgent'.

This book is concerned almost exclusively with Antarctica. But Antarctica's rise on the international political agenda has not occurred in isolation. Governments and citizens are now far more conscious of the need to protect the environment generally.

The Second Environmental Wave is underway. The First Environmental Wave occurred during the 1950s and 1960s at a time when industrialized nations became worried about the new levels of pollution. Concern about pollution was not new but the post-Second World War industrial expansion took pollution to unprecedented levels. This concern culminated in the 1972 Stockholm UN Conference on the Human Environment.[2] Environmental issues became embedded in national and international political agendas.

Antarctica itself did not figure in the First Environmental Wave. In 1959 the nations involved in Antarctic research agreed on a magnificent treaty which froze issues of potential dispute (notably over territorial claims) and provided a firm legal foundation for international scientific research. There was no cold war in Antarctica. Additionally, Antarctica was too far away from areas of industrial pollution and so the worries which troubled people elsewhere had little relevance for Antarctica. The first environmental wave did not, therefore, have much direct bearing on Antarctica – though the Antarctic Treaty's success may have provided some inspiration to the participants in the First Environmental Wave that international environmental diplomacy could, in fact, blossom.

Antarctica has a different role in the Second Environmental Wave. This book deals briefly with two elements of current environmental concern: the greenhouse effect (global warming); and the hole in the ozone layer over Antarctica. Other transnational environmental issues threatening the earth, such as acid rain, dioxins from paper-bleaching factories, waste disposal and the need for recycling, do not receive attention in this book.

The twin rise of both concern over Antarctica in particular and the environment in general has three implications for this book.

First, the new public concern about Antarctica has to be viewed against the larger canvas of public concern generally for the environment. One feeds into the other, such as the potential for rising sea-levels due to Antarctic ice melting because of the greenhouse effect.

Antarctica is here to stay on the international political agenda.

Second, Antarctica itself is now a matter of concern for many recent converts to the environmental cause. Two examples are the former British and Australian Prime Ministers. Mrs Thatcher, speaking in March 1989 on enhancing British research expenditure on Antarctica, said:

I had always been interested in Antarctica . . . You never quite know what's going on in those fascinating, remarkable icy lands. They are not wastelands . . . I did not know whether it was scientifically important. But I did think that it was very, very important for Britain. We have lands there. It was a fascinating place. We wanted to know more about the seas there, the wild life and the mineral deposits. And I did say the Antarctic Survey must have more money. And it was they who discovered the depletion of the ozone layer. . . . [3]

Meanwhile Bob Hawke (whose Government's remarkable transition on Antarctic affairs is examined in more detail in Chapter 4) said in July 1989:

> Antarctica is the world's last great wilderness. It provides a habitat for many living species. It also constitutes an extremely valuable scientific laboratory for measuring the extent and effects of global climate changes in the thickness of the ozone layer . . .
>
> But ultimately it is the sheer weight of international public opinion that will determine the future of Antarctica. Australia does not expect to meet with unqualified success overnight, but as international awareness of the hazards of mining in Antarctica continues to spread, more and more governments will be persuaded that mining in Antarctica is a risk that should not be countenanced.
>
> The Australian Government will continue to push its proposals vigorously: it will not relent in its efforts to prevent mining in Antarctica, to promote a comprehensive environment protection convention and to establish an Antarctic Wilderness Reserve.
>
> Australia's policy interests in Antarctica as agreed by successive governments since the 1960s have included the objective 'to derive any reasonable economic benefits from the living and non-living resources of the Antarctic'. Consistent with this decision not to allow mining in Antarctica, the Government has amended that policy objective to make it clear that it excludes deriving benefits from mining or oil-drilling.[4]

Third, protecting Antarctica is no longer just a matter of making part of the globe safer for scientists – it is now a matter of human self-interest. Potential rising sea-levels and a hole in the ozone layer mean that protecting the Antarctic equals protecting ourselves. Self-interest is always a potent force in politics.

A window of opportunity

This book is written from the perspective of advocating that Antarctica should be left as a wilderness, free from economic exploitation, political competition and military confrontation. The next few years will present challenges to the basic 1959 Antarctic Treaty but the immediate future may also provide a unique window of opportunity. Antarctica may well benefit from unprecedented levels of international political attention; much depends on how conservationists and politicians use that window of opportunity.

The background to the current problems confronting Antarctica is set out in this book. It also calls for Antarctica to be given a new and unique international status: the Public Heritage of Humankind.

The 1959 Antarctic Treaty was a bold innovation for the then international legal and political era of the 1960s. The way forward in the 1990s must be equally innovative for the current era.

Additionally, the Antarctic Treaty, although this may not have been recognized fully at the time, created a new trend: nuclear-weapon free zones. The Antarctic model has been followed in respect of the Moon and Outer Space (1967), Latin America (1967), the Sea-bed (1971) and the South Pacific (1986). This book sets out a new concept – the Public Heritage of Humankind – which may again enable Antarctica to exist as the catalyst for a new trend in international law and politics. Nuclear-weapon free zones led the international drive for arms control agreements – and the Antarctic Treaty set out the first such zone. Now the interest is in global interdependence and international co-operation to save the environment. Antarctica can again be used to help lead the way forward.

Finally, the 'Fall of the Wall' and the improvement in US–Soviet relations are all examples of a new international political climate. The patterns and obstacles of post-war politics are changing. Not all the changes are for the better (such as the resurgence of ethnic nationalism in eastern Europe) and Utopia is not about to descend to earth. But at least the old political deadlocks have eased, governments are more willing to listen to novel ideas, and the environmental crisis has captured the public's attention. Proposals which may have seemed too radical previously may now stand a greater chance of acceptance. Today's revolutionary suggestion may be tomorrow's conventional thinking.

We have no way of estimating just how long this window of opportunity will remain open. No time, then, should be wasted. A sense of urgency should pervade our work on protecting the environment in general and Antarctica in particular.

Notes

1. 'Fossil Find Backs Hot Earth Theory', *The West Australian*, 4 March 1989.

2. For a good introduction, see: Peter Stone *Did We Save the Earth at Stockholm? The People and Politics in the Conference on the Human Environment* London: Earth Island, 1973.

3. Margaret Thatcher, 'Nature Programme', BBC Television 2 March 1989, quoted in Peter J. Beck, 'A New Polar Factor in International Relations', *The World Today*, London, April 1989, p. 68.

4. R.J.L. Hawke, *Our Country, Our Future: Statement on the Environment*, Canberra, Australian Government Publishing Service, 1989, pp. 26-7.

2. The 1959 Antarctic Treaty: No Cold War at the South Pole

Good news is not news

The 1959 Antarctic Treaty has been one of the most successful treaties negotiated this century. It froze some difficult issues and created a base upon which the international scientific community could operate despite international political tensions on the rest of the globe. This record of success is one reason for the lack of attention, until recently, by the mass media to Antarctic affairs: good news is not news.

This chapter explains the content and significance of the 1959 Treaty. It begins with a brief introduction to what is international law and how nations 'acquire' territory. Antarctica is the world's last continent to be 'acquired' by nations and the conflicting claims to Antarctica are examined in the following section.

A breakthrough came with the International Geophysical Year, 1957–58, which showed that scientists could co-operate even if their political leaders could not. The example set by scientists was followed by governments and was enshrined in the 1959 Antarctic Treaty. This chapter ends with an examination of this remarkable document, and its main strengths and weaknesses.

What is international law?

> The Law of nations, or International Law, may be defined as the body of rules and principles of action which are binding upon civilized states in their relations with one another.[1]

International law as it is known today is essentially a European creation. Its roots go back to the rules developed by Roman and even earlier civilizations as a way of regulating their affairs. But the real growth of international law came after the Reformation and the growth of the European nation-states. These states were confronted with common problems which required common regulation, such as diplomacy, wars and shipping.

In thinking about international law, it is useful not to compare it with national law. The latter is mainly concerned with the regulation of individual behaviour (as far as criminal law is concerned) and this in turn means the existence of enforcement measures. International law, though it has some direct application to individuals (such as the capture of pirates and protection of foreign diplomats), is primarily concerned with regulating the behaviour of nations. International law does not have the same means of enforcement. The United Nations Charter provides for the creation of an international military force but this has not been created. There is the International Court of Justice (ICJ), but this does not 'punish' governments as such; indeed, many nations have refused even to appear before the ICJ (something the ordinary person accused of a crime cannot do).

International law does, however, have some measures of enforcement. If a nation acquires a reputation for breaking international law, then other nations will be wary of doing business with it. Also most international law obligations are not onerous and governments can follow them without risking their national security. Indeed, international law is followed more often than it is given credit for. Health regulations for overseas travel, the passage of foreign aircraft over a country, the posting of mail to other nations – these are but small illustrations of international law at work every day affecting our lives far more than most people realize.

Most international law is followed almost all the time. This is simply because it has been created to meet the specific needs of the nations concerned. International law is, therefore, on stronger ground than some national law, which is imposed on the community by the government. International law is, by contrast, built upon consensus. This makes the creation of a new law a lengthy process. But once created it is followed, simply because it accords with the wishes of most governments. If some aspect does not appeal to a government, then it is still rare for that government to disobey it; international law overall will be regarded by that government as too important to undermine, and so it would not want to undermine that particular aspect, in case its example is followed by other governments in other aspects. In this sense, then, international law creates a club, whose members (governments) have a mutual interest in ensuring that all the rules are kept, even though some may be contrary to their own particular views.

Where does international law come from?

There is no international 'parliament'. The United Nations is not a world government and the UN General Assembly is not a world parliament. Consequently, there is not one central source of international law. If a dispute arises between two or more nations, then the governments' international lawyers will have to refer to at least four sources of international law.

Treaties constitute the most important source. They come in two forms: bilateral (between two nations) and multilateral (between three or more nations). All the treaties covered in the rest of this book are multilateral and so there is no need to describe examples now.

A bilateral treaty arises when two nations wish to formalize an arrangement. Thousands of bilateral treaties have been created during this century. Almost all of them are so technical that they receive no coverage at all in the mass media – though they can mean a great deal to some people, such as the reciprocal payment of old-age pensions to people of one nationality living in another nation, and arrangements governing the exchange of students of one nationality to study in another nation. The most famous bilateral treaty of the late 1980s was the 1987 treaty between the US and USSR on the elimination of their intermediate-range and shorter-range missiles (the INF Treaty). This was the world's first nuclear weapons disarmament agreement and led to the scrapping of the land-based missiles in Europe which had done so much to generate fears of a 'limited' nuclear war.

The second source of international law comes from the customs of nations. In other words, nations have separately arrived at a similar practice. For example, pirates for at least 2,000 years have been seen as outlaws to be punished, usually by death. This custom evolved within each nation and, well before the UN came into existence in 1945 and began codifying and developing the law of the sea, there were already well established national laws against pirates. The strength of a customary rule, then, arises from the conduct of those whom it binds; it is followed simply because it is already being followed; nations accept as international law what they already regard as national law.

Customary international law used to be the main source of international law. But, since around the time of the UN's creation, it has been overshadowed by treaties. One reason for its comparative decline is its lack of specificity, or perhaps more accurately lack of agreement on the details of the customary rules. Piracy (to stick with one of the oldest examples) is outlawed but the application of that rule varied from one nation to another; Sir Francis Drake was a hero in England but regarded as a pirate in Spain. A second problem is that international law is seen as a European invention. Nations which now constitute the Third World were 'undiscovered' by Europe (and therefore 'unknown' to the 'civilized world') or were only European colonies at the time when many international customs were evolving. They had no participation in this process.

Treaty law, by contrast, is more specific. But there are still several 'gaps' in any treaty. Law is always a very blunt instrument, be it in international relations or within each nation, such as that covering crime, divorce and contracts. No law can cover every eventuality.

Additionally, the creation of a treaty gives greater scope for participation. The process of creating a multilateral treaty is usually long and thorough. But no nation – even if it were one of the treaty's principal architects – can be forced into accepting a treaty. Even the acceptance process can be prolonged; no nation need feel hurried. The first stage is signature: this is done by an authorized representative of a nation, signifying its intention, at a later date, to become bound by the treaty. The second stage is ratification: this is the formal consent by the nation to be bound on the international plane by that treaty and is done through the deposit of its instrument of ratification (in other words, when Australia ratifies the 1982 UN Law of the Sea Convention it will send to the UN Secretary General – the depository – a copy of the Act of the Australian Parliament bringing Australia's laws into line with the Convention).

A third source of international law arises from judicial precedents. Once again, it is necessary not to view international law through the prism of English or Australian law. International law follows the Continental European pattern whereby an international decision is binding only upon the parties to the case and in respect to that particular case. Precedents are not, then, binding authorities in international law – whereas they are, of course, in England and Australia. Additionally, international adjudications are rare – the International Court of Justice at The Hague and its predecessor the Permanent Court of International Justice have not been used by governments nearly as much as their architects hoped. (Governments prefer negotiation and mediation rather than a clear cut judicial decision in which one party will lose.) Judicial precedents are, then, of some value as guides but are nowhere as significant as the two previously mentioned sources of international law.

A fourth source of international law are writers on international law. They help clarify what the law is on a particular issue and what it ought to be. As the preceding paragraphs have demonstrated, international law can be very complicated because there is not one single source of it. International law scholars can review national practice, find areas requiring new treaties and may even provide draft treaties in their books and law journal articles. For example, Raphael Lemkin coined a term during the Second World War for the mass murder of an ethnic group: 'genocide'. He called for the punishment of the Nazi leaders for this crime (this was done at Nuremberg in 1945) and for there to be a multilateral treaty prohibiting genocide (the 1948 Genocide Convention). As Leo Kuper has commented, Lemkin's 'one-man crusade' was 'a remarkable achievement'.[2]

Another example with which most people are familiar – and which still has some application today in Australia – is the three-mile territorial sea. How far out to sea can a nation claim to control the water? That question was put to Cornelius van Bynkershoek (1673–1743). In 1737 he advised: three miles. That was roughly the maximum range of a cannon in the eighteenth century.

The acquisition of territory

The control of territory is one of the most fundamental concepts of international law. Without that territory, an entity cannot claim to be a nation. International law applies mainly to nations (or governments), not to individuals or non-governmental organizations (except in a minute number of instances). If an entity wishes to enjoy the rights and duties of international law, it must have territory. Once an entity has territory, it has room to place its population (assuming that the land did not have any previous inhabitants), it has control over the internal waters, the territorial sea just off its coast, the air space over the land, the possibility of membership of the UN, and countless other benefits as well as duties.

The acquisition of territory has been a matter of great interest to international lawyers because it does more for the territory than just providing the entry ticket to the club of nations. After all, if territory alone were the basic criterion, then the world would consist of small packets of territory, since there is no limit to the size of a nation; nations vary in size from the Soviet Union down to islands in the Pacific smaller than the area of London.

The problem of the acquisition of territory has been derived from the old belief that the more territory one can control the better. Rulers have not been satisfied with owning a small piece of territory; they have often sought to gain additional territory, thereby adding to their market for products, potential recruits for their armed forces, the desire to see the 'map painted red' (such as the British Empire).

There have been various methods for acquiring territory. *Occupation* means the settlement of a territory which was previously not controlled by another European power. This is a rather blunt way of putting it, but it reflects the underlying history of international law as a European institution serving European ends. Thus European colonizers travelled the world and occupied 'vacant' land for their rulers. This method is no longer as widespread as it was as there is no additional land – except for the unusual case of Antarctica – to occupy. *Accretion* is a minor method of acquiring territory and is derived from the gradual deposit of soil by a river flowing past a shore or by an ocean along its coasts. *Prescription* means continued occupation, over a long period of time, by one state of another state's territory, virtually by 'squatting' on it. The period involved is still in some doubt. There have been some cases of this, but it, too, is rare. *Voluntary cession* means the formal transfer of control of territory from one nation to another. This may be done as a sale or it may be an outright gift. This, too, is now rare. *Involuntary cession by conquest* is, with occupation, the main way of acquiring control over additional territory. It differs from occupation in that the conquest has to be of an independent member of the community of nations; thus African land was 'occupied' in the last century, whereas the Italian subjugation of Ethiopia in 1936 was 'involuntary cession by conquest'.

From that brief account of acquisition of territory, it is evident that we are approaching the end of the era of such acquisitions. First, the major way of acquiring new territory – by conquest of previously independent states – is now outlawed by the UN Charter, whose Article 2(4) prohibits the use or threat of use of force in international relations. This, unfortunately, does not stop the continued use of force, but most instances are now of internal conflicts rather than international conflicts; the former are not outlawed by the UN Charter, but the latter are.

Second, one of the revolutions in the post-Second World War era has been the demise of most empires. The trend, in other words, is not to acquire territory but to support national liberation movements. Most wars of national liberation have been won.

Third, the formal control of someone else's territory is now avoided since some governments believe that they can achieve the same results – additional power – by alternative methods. This new form of 'colonialism' has resulted in nations remaining independent in name, but with foreign nations having a great say in how they function, for example, the United States in Latin America and (until recently) the Soviet Union in eastern Europe. These countries maintain their influence through such mechanisms as economic power and the CIA and KGB. Nonetheless, they still claim not to have 'acquired' territory.

Fourth, there have been the post-Second World War attempts to 'internationalize' previously unclaimed areas. The 19th century scramble for Africa was held partly responsible for the subsequent wars in Europe and it has been hoped that by internationalizing certain areas – notably outer space and the sea-bed – these areas will not become the cause of future conflict since no nation's claim to control would be accepted by other nations. This work has been done via the concept of the common heritage of humankind.

Finally, there has been the growth, since the Second World War, of the recognition of the need for wealthy nations to help the developing ones. The help provided has not been adequate. But there is at least some acceptance of the principle that all nations should benefit from the world's resources, such as the proposed exploitation of the sea-bed.

To sum up, there is now a clear trend away from acquiring additional territory towards the granting of independence to colonies, as well as towards the internationalization of previously unclaimed land, such as the sea-bed.[3]

Antarctica remains the main exception to all these developments.

National claims to Antarctica

The Antarctic is the most inhospitable place on earth. It suffers the lowest natural temperatures ever recorded, and for the four months of winter it is totally dark. Gales blow for some 300 days a year. The region is isolated from the rest of the world by the Southern Ocean, by icebergs, and by fogs, all of which deter regular visits except by some fishing vessels and scientists. The summer population of the Antarctic is of the order of 4,000, with only 800 (mostly scientists) staying for the winter. Apart from some pollution carried in by the sea and air from distant lands, the continent is free from pollution. It is too cold for diseases to thrive. The continent consists of rock and ice. The rock is almost entirely buried under fresh-water layers of snow and ice and is visible in only a few places, mainly in the mountain ranges. The water surrounding the Antarctic is, however, so teeming with life that it is probably the area of the world's richest marine fauna. The marine life is cyclical. The southern seas are constantly enriched by nutrients from warm latitudes caught in the immense circulatory systems governed by winds and temperature changes, ultimately, of course, by the sun and the movement of the earth on its axis.

To reach Antarctica it is necessary to travel over the inhospitable Southern Ocean for some thousands of miles. Once the mariner has penetrated the Ocean and the fog, there is the further obstacle of floating pack-ice, whose location and depth are always changing. It is not difficult to understand, then, why Antarctica has for so long remained outside the scramble for territory which has been such a prominent feature of the world's history for the past few thousand years.

There have been three main motivations for colonization elsewhere in the world. First, there was the economic motivation: to gain raw materials or additional markets. Second, there was the strategic motivation: the belief that it was necessary to hold on to certain territory as a launching point for other invasions or to have a defensive or protective belt from invasions. Third, there was a mixture of psychological motivations: the need to win over 'heathens' – to make them good Christians or Muslims – or to win greater glory for one's nation by expanding the size of its empire. All three motivations did not apply to Antarctica because it could not, and to a lesser extent, still cannot, satisfy any of these motivations.

The first people to encounter physical proof of the Antarctic were probably Polynesian canoeists travelling from South Pacific islands. These people sailed south until stopped by ice, and they contributed to their national culture legends about a great white land to the south. European mariners from the fifteenth century onwards navigated the South Atlantic and Indian Oceans, but no definite progress was made until Captain Cook's three voyages between 1768 and 1775. Even then the indications were unclear – the British mariner reported that if any land did exist there it would be inhospitable. Around 1819–20 European and American mariners working in the whaling and fur seal industries ventured close to it. There is still some controversy regarding who reached there first but it was probably the Russian, Thaddeus Bellingshausen.

Although the next century or so was notable for continual navigation work (including claims by various people to be the first to set foot on the continent), the principal mariners in the area remained whalers and sealers. The beginning of the 20th century saw the so-called heroic age of exploration, including the race to the South Pole. Around this time, some explorers started making claims to Antarctic territory.

National claims to Antarctica have been pressed less fervently than claims to territory elsewhere. Indeed, the first claim (by the UK in 1908) was so haphazardly done that it took in part of South America and had to be reframed in 1917. Moreover, Antarctica is also unusual in that, as already noted, for much of this century three claims (by Chile, the UK and Argentina) have overlapped. But, uniquely, this competition has not given rise to any war or any attempt to have the claims arbitrated; in fact some nations have refused such arbitration.

To add to the complexity, some claims are not recognized by other nations; the United States and Soviet Union have consistently denied the validity of any title to Antarctic territory, and have treated the whole continent as open to their activities. Both, however, have reserved their 'rights' to make claims – though neither has given any indication of doing so. (There is still a segment of Antarctica that remains unclaimed.)

In short, Antarctica represents a different problem in terms of the international law relating to the acquisition of territory, since practices of international law which have developed for other parts of the world have had little application here. For example, an important criterion of 'control' of claimed territory is that there is a physical presence of the claimant nation in the territory – but how does a nation establish a permanent presence in such an inhospitable continent? In other words, it is not enough for an explorer to place a national flag on an 'unclaimed' territory (that is, traditionally one that was not already governed by a European nation) and claim it for the explorer's nation;[4] there has to be effective occupation. Ironically, in terms of establishing a 'presence' the US and USSR each 'occupys' Antarctica more effectively than any of the seven claimant nations.

After the Second World War, there was an American initiative to try to settle the differing claims. The need for such an initiative was made clear during the late 1940s and early 1950s by the acrimony that developed between three countries whose claims overlap. Argentina, which has long opposed the UK's claims to the Malvinas/Falkland Islands, temporarily overlooked its own quarrel with Chile to work for the recognition of the 'South American Antarctic' in opposition to the UK. Despite a 1948 agreement among the nations which was aimed at maintaining a fragile truce, there were later incidents involving gunfire, although no one was killed. On 4 May 1955, the UK instituted proceedings before the International Court of Justice against Argentina and Chile concerning the disputes as to the sovereignty over certain lands and islands in the Antarctic. Unfortunately both Chile and Argentina had not accepted the Court's jurisdiction and so on 16 March 1956, the Court decided not to proceed with the question. Had Chile and Argentina co-operated with the Court, it might have been able to throw additional light on the vexed problems of claims to Antarctica. Nonetheless, the incidents did augment the growing belief that some agreement had to be reached to facilitate the use of the continent.[5]

International Geophysical Year

> It was Alexander von Humboldt who, in the early nineteenth century, did most to revive the medieval concept of a commonwealth of science; for he managed to persuade the British and Russian governments to set up joint stations all over the world from which to conduct the study of geomagnetism. His lead was followed by the German Gauss, the American Maury and the Austrian Weyprecht; and it was largely due to the pioneering work of these far-sighted scientists that the first International Polar Year was held in 1882. A second Polar Year was held in 1932–33, and the idea of holding a third was mooted soon after World War II by a group of US scientists led by Dr Van Allen; 1957 was suggested as a suitable year, since sunspot activity was then expected to be at a maximum. The task of getting such a scheme started was daunting in the extreme, for the Cold War was at its zenith, and Antarctica itself was bedevilled by conflicting and often bitterly contested claims. It soon became apparent, however, that the scientists of the world were better ambassadors for peace than the politicians; and, to quote H.G.R. King, 'the spirit of mutual trust and co-operation displayed in the early stages of the IGY has characterised the relationships between all nations in Antarctica from that day to this'.[6]

The International Geographical Year was one of the special 'years' that the UN has since used so often, for example, 1986 was the International Year of Peace and 1990 is the International Year of Literacy; the 1970s and 1980s were known as the Decades for Development and Disarmament. The IGY was a world-wide co-ordinated study of natural phenomena relating to the world as a whole. It lasted for an 18-month period, 1 July 1957 to 31 December 1958. The IGY was the most complex and comprehensive international scientific activity ever undertaken. By the time the Year was underway in 1957, some 50 stations were staffed in Antarctica by scientists from twelve countries with programmes in the region, while worldwide, over 10,000 scientists and technicians from an eventual 67 participating countries worked at 2,500 stations by the time IGY ended in 1958.

The IGY survived the political turmoils of the cold war. It is clear, however, that some political rivalry entered into the IGY Antarctic operations, especially between the United States and the Soviet Union. The Soviet Union had not taken much interest in the pre-Second World War Antarctic developments, but reversed its policy after the war. The Soviet Union was particularly active in the IGY, and this led to the American decision to double the original number of projected US bases. The Soviet Union, which had unexpectedly agreed to allow a Swiss expedition to operate in the Soviet Arctic, was anxious that the IGY friendship should also be realized in Antarctica. Relations between the Soviet Union and Australia were particularly bad at this time (in strict diplomatic terms, they were non-existent) because of a Soviet spy scandal in Australia. Nonetheless, Soviet and Australian scientists got along well. Just prior to the IGY, the Soviet team decided to celebrate the Soviet Communist Party Congress by hoisting the Soviet flag to the boom of guns and the national anthem. The raising of the Soviet flag over the Australian Antarctic Territory caused a sensation in Canberra among politicians and mass media; but the scientists in Antarctica took it quite calmly.

The Antarctic's contribution to IGY was generally regarded as a major success in scientific and political terms. This was all the more so since the prevailing 'cold war' elsewhere meant that Western–Communist relations were warmer in Antarctica than elsewhere around the globe.

> It was inconceivable that all this arduous, expensive and meaningful work should suddenly come to an end on December 31, 1958. And indeed before the IGY was even under way a committee had been set up to examine how scientific research in Antarctica might be continued. This committee (SCAR) had one overriding objective: to nurture the seed of international co-operation. It did valuable work. It recommended the continuation of a wide range of investigations – meteorology, atmospheric radiation, auroral

physics, geomagnetism, geology, glaciology, geomorphology, cartography, seismology, vulcanology, oceanography, biology and marine biology; it perfected a system whereby all findings from these sciences were automatically fed back to a number of International Data Centres and thence disseminated throughout the world; it also helped to bring about the climate of goodwill and co-operation that led to the signing a few years later of one of the greatest documents of modern times: the Antarctic Treaty.[7]

The 1959 Treaty

The success of the International Geophysical Year encouraged hopes of making the spirit of scientific co-operation more permanent. The 1959 Antarctic Treaty is a remarkable agreement. It was signed on 1 December 1959 in Washington DC, by representatives of Argentina, Australia, Belgium, Chile, France, Japan, New Zealand, Norway, South Africa, the Soviet Union, the United Kingdom and the United States of America. These were the original twelve 'Consultative Parties', who had been involved in various IGY activities in the Antarctic. The Treaty came into force on 23 June 1961 and consists of 14 Articles (see Appendix 1 for the full text of the Treaty).

Article I states that Antarctica shall be used for peaceful purposes only and the contracting parties (those countries which agree to be bound by the Treaty) agree not to establish any military bases in the area, or to carry out military manoeuvres or weapon-testing.

Article II provides for freedom of scientific investigation, and co-operation towards that end shall be maintained.

Article III provides for scientific information and personnel to be exchanged by the contracting parties.

Under Article IV nothing contained in the Treaty may be interpreted as a renunciation, denial, or support of a claim to territorial sovereignty in Antarctica. No new claim to territorial sovereignty may be asserted while the Treaty is in force.

Article V states that the Antarctic may not be used for nuclear explosions or the disposal of nuclear waste. However, this Article does not prohibit the creation of international agreements concerning the peaceful use of nuclear energy in Antarctica by the contracting parties.

Under Article VI the Treaty's provisions apply to the area south of the 60 degree Latitude South including all ice shelves. But the Treaty does not prejudice the rights of any nation under international law with regard to the high seas within that area.

Under Article VII each contracting party has the right to send observers to carry out inspections in Antarctica. Notifications must be given to all the contracting parties of impending expeditions. Under Article VIII observers and scientific personnel are subject only to the jurisdiction of their own nations.

Under Article IX, the contracting parties shall meet at suitable intervals to consult together on measures to further the Treaty's aims, and under Article X, contracting parties agree to exert appropriate efforts, consistent with the Charter, to the end that no one engages in any activity in Antarctica contrary to the Treaty's purposes. This article is so worded to apply to all humans, whether governmental or non-governmental, and not just the contracting parties.

Article XI requires contracting parties to settle their disputes by peaceful means, including the International Court of Justice (providing the parties concerned all agree to refer their disputes to the ICJ).

Article XII provides for amendments to the Treaty. Also, when 30 years have elapsed since the Treaty entered into force (after 23 June 1991), the Treaty may be reviewed if any of the contracting parties so desire. But there is no automatic obligation in 1991 to review the Treaty.

Article XIII contains the standard references to the coming into force of treaties; and Article XIV concerns the usual matter of the equal authenticity of the four versions (English, French, Russian and Spanish) of the Treaty's language.

An assessment of the Treaty

The Treaty has eight main strengths. First, it was drafted at a very difficult time in international relations, because of the cold war, and so it is amazing that anything was written at all. Additionally, it reflected US dominance in world affairs – and the USSR's pragmatic acceptance of that dominance in Antarctic affairs. The US had decided not to press its territorial claims (unofficially established before the Second World War by US Admiral Richard Evelyn Byrd) to Marie Byrd Land (named after Byrd's wife). Instead, it conceived of an international condominium, in which national claims would be held in abeyance, but where it would play a leading role. Since American power was then unchallenged, this was the view that prevailed. It was enshrined in the Antarctic Treaty: an elegant mixture of US self-interest and international high-mindedness.

Second, the Treaty is the first international agreement which, being a treaty establishing a demilitarized zone, contained by implication provisions to ensure that no nuclear weapons would be introduced into that area. Though perhaps only a minor breakthrough in itself, in terms of limiting the arms race, it has been a helpful precedent for more ambitious nuclear-free zones, notably Latin America, the Moon, the Sea-bed and South Pacific. It also helped the momentum for international arms control negotiations. It is worth noting that until the 1987 Intermediate Nuclear Forces Agreement between the US and USSR, the USSR had always opposed all forms of international inspection in disarmament and arms control treaties in so far as they applied to Soviet territory. Right from the outset of the Antarctic Treaty, however, the USSR accepted inspection of its Antarctic bases. Australia, for example, carries out at least one inspection each summer of the bases of other nations (including the USSR's) in and around the Australian Antarctic Territory. This inspection procedure has been (to use the current jargon) a highly effective 'confidence building measure'. It is possible that the USSR's experience in Antarctica has contributed to the easing of its hostility towards such inspections in recent disarmament negotiations. Additionally, this 'zone of peace' towards Australia's and New Zealand's south has meant that both have not had to worry about defence planning for their southern flanks.

Third, most of the Antarctic is claimed by Australia, Norway, Argentina, Chile, the UK, New Zealand and France. Marie Byrd Land is not claimed. The claims are disputed. Other nations emphasize that they recognize no claims at all. (Australia's claim is recognized only by the UK, New Zealand, France and Norway.) The Treaty has sidestepped this problem by Article IV's statement on, in effect, freezing the claims issue.

Fourth, the Treaty has also avoided embarrassing political divisions over the claims since all the claimants are, to varying extents, identified with the same superpower – the US – and they tend to have common regional political outlooks: UK – France – Norway; Australia – New Zealand; and Argentina – Chile. Antarctica has not generated intra-alliance tensions. Thus, for example, the Icelandic Cod Wars with the UK, and the Greek/Turkish dispute over Cyprus, put great strains on NATO, if only by implying that local issues are more important than the need for unity to oppose 'communist expansion'. By contrast, Articles I and IV have kept Antarctica out of armed confrontations.

Fifth, the Treaty is unusual in that it has created some international machinery. Many nations are willing to accept international obligations (via treaties), yet want to reserve for themselves the obligation of ensuring that they keep their obligations. This concern is derived from a reluctance by almost all nations to have any international interference in their domestic affairs. This Treaty, by its consultative arrangements and mutual obligations, has created a limited international system of enforcement.

Sixth, the test of any agreement is whether it is fulfilling its original aims; this Treaty certainly is. Memoirs of scientists who have worked in Antarctica during the last three decades attest to international co-operation. Meanwhile, the consulting parties include a number of argumentative nations which at public international gatherings, such as UN conferences, devote some time to abusing each other: notably the three Communist nations – the Soviet Union, Poland and East Germany – and South Africa and Chile. It is notable that the latter two nations now have no diplomatic representations in the three Communist countries and vice versa (although the rapidity of events in South Africa and eastern Europe may soon change this situation): Antarctica, however, brings them together away from the glare of international publicity. Similarly, the UK and Argentina during the 1982 Falklands/Malvinas conflict maintained tranquil relations over Antarctic affairs.

Seventh, the Treaty has provided a firm foundation for flexible growth. The twelve original signatories have been joined by 27 other nations. The 39 nations represent a variety of the world's political and economic systems; their populations account for over three-quarters of the world's population; the newer members have been absorbed with minimal disruption (except for the customary proliferation of paper work and lengthier negotiations); the consensus decision-making system has remained intact. Additionally, the Treaty did not address questions such as pollution control, tourism, political status, living and non-living resources. It aimed for – and achieved – the highest common area of agreement on matters requiring urgent decisions, while not excluding subsequent developments to fill in the gaps.

Finally, the Treaty is a show-piece of international law in operation. This Treaty is self-enforcing. It is in each nation's interest not to violate it, because the gains from respecting the Treaty's provisions far outweigh the gains that might result from each nation attempting to militarize its own zone.

Limitations of the Treaty

The Treaty is, then, a useful document, but it is possible to overestimate its significance. First, it has created an elite minority of 25 nations (the Consultative Parties). Three decades ago such a creation was possible because the largely European nations and the US were accustomed to running the world and many of today's Third World nations were still colonies. It is reasonable to assume that if the Antarctic Treaty were written now, it would have to make alternative arrangements for the Consultative Parties, in the sense that more nations would want full rights of participation. Indeed, it might well have to take a totally different form, to take into consideration the growing interest in the concept of the common heritage of humankind; this point is returned to later in this book. Additionally, the Consultative Party core contains some anomalies. South Africa's participation is examined in Chapter 5. Belgium has no Antarctic involvement today in terms of scientific research – a hurdle to be overcome by all of today's aspiring Consultative Parties.

The Treaty's membership situation (as at June 1990) is set out in the Annex to this chapter. Since this is one of the main controversies at the annual UN General Assembly it is worth looking at this matter in extra detail.

The 39 nations which have ratified the Treaty fall into three categories. First, there are the original twelve nations which were active in Antarctic scientific work in IGY: the first Consultative Parties. Second, there are now 13 other nations which also have Consultative Party status. These nations govern Antarctic affairs.

Any nation – and there are approximately 170 around the world – may ratify the Treaty. But ratification does not guarantee automatic acceptance to Consultative Party status. The third category, then, consists of nations which have ratified the Treaty but do not have Consultative Party status. The Consultative Parties alone determine which nations may become Consultative Parties. A key requirement is scientific activity in Antarctica. There is no set period as to how long a nation takes before making it into the inner circle. To take the examples of the three most recent Consultative Parties (1989), Finland ratified the Treaty in 1984, Peru in 1981 and South Korea in 1986. By contrast, India – which is a key Third World nation in the debate over Antarctica – ratified the Treaty on 19 August 1983 and became a Consultative Party on 12 September 1983. As with an exclusive club, the admission of new members partly dilutes the privileged status of the existing members. These must balance off the advantages of admitting new members with the disadvantages of diluting some of their own authority. India's inclusion was a shrewd move – as will be shown in Chapter 5.

If becoming a Consultative Party can be a problem, ceasing to be one is even more difficult. The Treaty contains no provision for denunciation (that is, a nation deciding to cease to be bound by it) and no provision for expelling a party (Consultative or not). As previously noted, Belgium, which was one of the original twelve Consultative Parties, would not be eligible for that status nowadays because it no longer conducts Antarctic research. But it cannot resign or be expelled.

A second limitation is that while no nation has made additional claims in the Antarctic, no nation has renounced any of its claims. Chile and Argentina, in particular, have sought to gain increased international recognition of their claims by such symbolic activities as holding national ceremonies there. Moreover, a cynic may argue that one of the main motivations for all the nations which claim to conduct scientific research there is that the research reinforces the notion that they do have claims. After all, Chile and Argentina do not otherwise figure prominently in international research conducted, for example, into marine biology.

Third, the Treaty did not create an international organization to govern Antarctica. There is not, for example, a standard currency (although the US$ is the most popular one), or a common set of rules for navigation or handling accidents. The research teams operate very much as national entities. Indeed, it is worth recalling that Article VIII(1) of the Treaty states explicitly that observers and scientific personnel remain under their respective national jurisdictions. They may co-operate and establish friendly relations, but they are not an international community of scholars; they are not like the people from various nations employed to work, for example, as UN civil servants.

Fourth, the Treaty is almost as significant for what it omits as for what it contains. In recent years there has been growing international interest in the exploration and exploitation of the living and non-living resources of Antarctica. Apart from fisheries, which are believed to be rich, it is the prospect of mineral deposits (especially off-shore oil) which is attracting this new interest. But the Treaty makes no provision for this situation. The Treaty makes provision (Article IX(1)) for separate agreements for the preservation and conservation of living resources, or the limitations to environmental damage arising out of, for example, oil drilling. Living and non-living resources are the object of more recent treaties.

Finally, the Treaty is showing signs of its age in terms of contemporary standards of environmental protection. Back in 1959, national environmental legislation was rudimentary and 'ministries of the environment' non-existent. The 1959 Treaty is a product of that era: it contains no environmental protection agency and no infractions committee. If the Treaty were written today, it would pay far more attention to environmental protection. Indeed, some attention is now being given by the Consultative Parties to this matter (this will be dealt with later in this book).

A visionary treaty

It would be a pity to end this chapter on a negative note. The Treaty – though this may not have been recognized fully at the time – created new ways of thinking, such as the idea of a nuclear-weapon free zone. Such zones are now taken for granted and constitute a standard item of the arms control agenda. But at the end of the 1950s the UN disarmament negotiations were stalled. Various ambitious proposals were in circulation but they were not receiving governmental support – indeed, the world is still a long way from multilateral disarmament under international inspection. The Treaty, therefore, established a new way of trying to break the deadlock. It was the leading edge of arms control treaties. 'Disarmament' means the abolition of entire weapons; 'arms control' is less ambitious in that it seeks to slow the rate at which the arms race is moving, such as by establishing demilitarized zones. In the bleak years of superpower tensions when disarmament negotiations were stalled, such agreements helped efforts by concerned bodies to slow the arms race.

The Treaty's main impact was its use of vision. As later chapters will demonstrate, new crises are confronting the globe. The challenge, then, is to see if the Antarctic's legal and political situation can once again be used not only to help protect Antarctica itself but also to create new precedents – this time in protecting the global environment via international co-operation and a recognition of the globe's interdependence.

In other words, in the late 1950s/early 1960s the Antarctic Treaty helped ease the disarmament deadlock by offering a new route via arms control. Can the Antarctic be used to stimulate new guidelines in global environmental protection? I believe that it can. Before dealing with that challenge, however, it is necessary to look at how the Consultative Parties have tried to build upon the 1959 Treaty.

Notes

1. J.L. Brierly, *The Law of Nations*, Oxford, Oxford University Press, 1963, p. 1.
2. Leo Kuper, *Genocide*, London, Penguin, 1981, p. 22.
3. The best introduction to this is: Clyde Sanger, *Ordering the Oceans: The Making of the Law of the Sea*, London, Zed Books, 1987.
4. The most bizarre manifestation of territorial claim-making by planting a flag was done on Hitler's instructions. In 1938, Nazi Germany sent flying boats to Antarctica to drop swastika flags tied to javelins every 50 miles across the continent. But this was a risky operation – the javelins were heavy, the weather inclement and the aircraft fragile. It is possible that the pilots dumped the whole load in Norway's claim and returned to their support ships; the swastikas are somewhere in the ice sheet slowly being carried across the continent to the sea. See: Michael Parfit *South Light: A Journey to Antarctica*, London, Bloomsbury, 1988, pp. 170-171.
5. For some recent studies of these issues, see: F.M. Auburn, *Antarctic*

Law and Politics, London, Hurst, 1982; Peter J. Beck, *The International Politics of Antarctica*, London, Croom Helm, 1986; M.J. Peterson, *Managing the Frozen South: The Creation and Evolution of the Antarctic Treaty System*, Berkeley, University of California Press, 1988; Deborah Shapley, *The Seventh Continent: Antarctica in a Resource Age*, Washington DC, Resources for the Future, 1986; Gillian D. Triggs (Editor), *The Antarctic Treaty Regime: Law, Environment and Resources*, Cambridge, Cambridge University Press, 1987; Gillian D. Triggs, *International Law and Australian Sovereignty in Antarctica*, Sydney, Legal Books, 1986.

6. Ian Cameron, *Antarctica*, London, Cassell, 1974, p. 238.
7. Ibid., pp. 239-40.

Annex to Chapter 2

Antarctic Treaty Nations

Claimants	Antarctic Treaty Consultative Parties (ATCPs)	Non-Consultative Parties (NCPs)
These countries claim sovereignty over parts of the continent. (Chile, Argentina, United Kingdom claims overlap)	*Made up of claimants and non-claimants. ATCPs have full voting rights in the Antarctic Treaty System*	*NCPs conduct scientific experiments or have an interest in Antarctica, but have no voting rights*
Argentina	(i) Original Parties:	Austria (1987)
Australia		Bulgaria (1978)
Chile	Argentina	Canada (1988)
France	Australia	Colombia (1989)
New Zealand	Belgium	Cuba (1984)
Norway	Chile	Czechoslovakia (1962)
United Kingdom	France	Denmark (1965)
	Japan	Ecuador (1987)
	New Zealand	Greece (1987)
	Norway	Hungary (1984)
	South Africa	Netherlands (1967)
	United Kingdom	North Korea (1987)
	United States	Papua New Guinea (1981)
	USSR	Romania (1971)
	(ii) Later Parties:	
	Poland (1977)	
	West Germany (1981)	
	Brazil (1983)	
	India (1983)	
	China (1985)	
	Uruguay (1985)	
	East Germany (1987)	
	Italy (1987)	
	Spain (1988)	
	Sweden (1988)	
	Finland (1989)	
	Peru (1989)	
	South Korea (1989)	

3. Antarctic Living Resources

Whales

The 1959 Antarctic Treaty was written at a time when there seemed little prospect of the commercial exploitation of Antarctica's living and mineral resources. The Treaty's architects sought agreement on what could be agreed upon easily (relatively speaking) and so did not waste time on addressing hypothetical issues. Besides, whatever resources existed in Antarctica could more easily be acquired from elsewhere around the globe (not least from some of the Consultative Parties themselves).

In the context of living resources, however, two exceptions have presented ill omens for the systematic regulation of Antarctica's resources: whales and seals. Whaling has not been a matter for the Consultative Parties as such since this activity is regulated by the International Whaling Commission (IWC) (which, coincidentally, has an overlapping membership with the Antarctic Treaty).

Commercial whaling has been conducted for several hundred years by Europeans. It was, for example, the first industry in Australia following the creation of the colony in 1788. Whaling was also the first industry in Antarctic waters.

Antarctic whaling developed because other whaling areas were over-exploited. Indeed, it was the prospect of whaling in distant Antarctica that helped keep the ailing industry alive. But Antarctica's whales were larger and faster than those found elsewhere. Ordinary harpooning was ineffective and dead whales sank beneath the waves. Technology provided the answer in 1865: a gun was invented that fired an explosive-tipped harpoon to riddle the whale with shrapnel while making it stick into the whale's flesh with heavy steel barbs. Steam-powered ships enabled whalers to chase their victims. Carcasses were pumped with compressed air to keep them afloat. The industrial revolution was mobilized in the interests of whaling.

The extent of the problems associated with tracking, killing and shipping whales is difficult to comprehend. The first peace-time whaling season after the Second World War was not considered a good year by whalers. The catch consisted of only 8,865 blue and 12,877 fin whales. But by weight of flesh, the 1946–47 season resulted in the rough equivalent of 22,600,000 men – the number of soldiers of all nations killed in the Second World War.[1]

The history of whaling worldwide is one of greed working itself out of business. Whalers sought to maximize immediate self-gain and so destroyed the long-term basis of their industry. Attention was focused on improving the technology of death – rather than guaranteeing the industry's future by limiting the numbers caught.

International attempts to regulate whaling were, then, slow to commence. Although Antarctica was a major whaling area, most attempts to regulate whale catches have not focused on Antarctica alone. The slow evolution in this matter has thus been in terms of global whaling rather than specifically concerning Antarctica.

It was only in 1931 that the first whaling treaty was created.[2] It prohibited the commercial hunting of two particular species – right whales and bowhead whales. It also tried to prevent the killing of calves or suckling whales, immature and female whales which were accompanied by calves or sucklings. Despite its limited objectives, five whaling nations – Japan, Germany, Chile, Argentina and the USSR – refused to accept it.

Fifteen governments drew up in 1946 a convention designed 'to provide for the proper conservation of whale stocks and thus make possible the orderly development of the whaling industry'. The International Convention for the Regulation of Whaling entered into force on 10 November 1948. It established the International Whaling Commission, on which each member nation is represented by a commissioner, to manage the whale stocks. Its duties are to organize both the conservation and the utilization of whale resources. Its major decisions are binding on member nations, except that if a government disagreed, it could within 90 days lodge an objection, in which case the decision would not be binding on that government. This provision – similar to the veto in the Security Council – was insisted upon by the United States and some others.

The International Whaling Commission, therefore, began mainly with the task of killing whales; conservation was done only in the interests of allowing whale herds to recover sufficiently so that they could be killed later. The Commission was out to safeguard the industry's future – not that of the whales. The treaty never envisaged the total banning of all whaling.

The Commission's philosophy has, however, been changed – largely due to pressure from environmental non-governmental organizations (NGOs). Australia – the last commercial whaling nation in the

English-speaking world – changed its policy from permitting whaling to making it, in 1981, a criminal offence for any Australian anywhere in the world to indulge in commercial whaling.[3] New Zealand, one of the treaty's architects in 1946, resigned from the Commission in 1968, partly because it disapproved of the excessive whaling being permitted by the Commission. Owing to NGO pressure it rejoined in 1976 to work with other nations to stop whaling. The US, which initially had created the Commission's veto, also changed its attitude during the 1970s.

After years of international non-governmental and governmental campaigning an international moratorium was enacted in 1985 by the Commission. The Commission recognized that commercial whaling was threatening the continued survival of some species. Japan and five other nations have, however, pursued 'scientific whaling' for the purposes, they claim, of conducting research.

The 38 member-nation Commission meets each year. At the June 1990 session, the Commission estimated that for the 1987–88 season, Japan took 1,082 whales, Norway 375, Iceland 100, the Soviet Union (Alaskan Eskimos) 81, and St Vincent and the Grenadines two. The toll in 1990 could be higher. The Soviet Union, eager for the hard currency from exports to Japan, has announced it will resume limited whaling, after supposedly getting out at the end of 1987, and Japan has signalled its intention to increase its quota.

Australia, the UK, the US and the Netherlands are leading a campaign to stop them; they have the numbers. Since non-whaling nations joined the IWC in the early 1980s, it has turned from being a whalers' club that set the highest quotas (60,000 a year) during the 1960s, to becoming an anti-whaling lobby. Japan did not, in June 1990, achieve the three-quarters majority required to scrap the moratorium. But neither, it seems, will the IWC manage to stop Japan from hunting whales.[4]

This whaling is conducted in the face of mounting international opposition. This has included direct action by environmental groups to harass whalers at sea and the US's banning of Japanese fishing vessels from its waters. As many as 300 whales have been taken by Japan each year in Antarctic waters for 'research' purposes. International opposition is not going to fade away, but Japan is equally obstinate. There is some hope because Japanese fishermen apparently want their government to stop whaling so that they can regain access to US waters.

Another source of hope comes from the apparent effectiveness of the international campaign to stop the Japanese 'strip mining' the sea. Drift net fishermen set fine-meshed nylon nets up to 60 kilometres long and 15 metres deep, trapping almost everything in a massive swathe of ocean. This fishing method is now widely used because it lands massive hauls of fish – despite spoiling up to 40 per cent of the catch. It is used in the South Pacific to catch albacore tuna. Conservationists are outraged because driftnetting destroys commercial fish stocks and ensnares whales, dolphins, turtles, sea birds and other marine life.

The nets are suspended vertically in the water by floats at the top and weights at the bottom. They are so sheer, they are virtually invisible. Even dolphin sonar does not detect them and whole schools of dolphins and small whales are entrapped. Diving birds are entangled and reportedly die by the hundreds of thousands. Nets are damaged, lost and abandoned, and these non-biodegradable 'ghost nets' are left to entangle and kill marine animals, birds and fish forever.

Ironically, the Japanese – who 'strip mine' the South Pacific high seas – do not permit driftnetting within 1,000 kilometres of their own coast.

On 18 July 1990 Japan announced it would suspend driftnet fishing by its fleets in the South Pacific from October 1990. The move was welcomed by South Pacific governments and was likely to improve Japan's image in the region where it had been making a concerted diplomatic push. A statement from the Japanese Embassy in Canberra said the suspension of driftnet fishing would continue until such time as appropriate conservation and management arrangements for South Pacific resources are devised. Japan will not resume driftnet fishing unless there is a consensus in favour of it among the South Pacific nations. As all South Pacific nations are committed to the total banning of driftnet fishing, this is not likely. Although the Japanese move means driftnet fishing is likely to cease altogether in the South Pacific, Japan, South Korea and Taiwan maintain about 1,000 driftnet fishing ships in the northern Pacific.[5]

The driftnet saga in the South Pacific shows that it is possible to influence even the most obstinate of nations. But the campaign to do so was time-consuming and only partially successful (because it will continue in the northern Pacific).

To conclude, the story of Antarctic whaling has five lessons. First, greed often triumphs over common sense. Whalers went out for maximum short-term gains – even though their actions jeopardized their long-term employment prospects.

Second, governments and whalers were slow to heed the warnings of those scientists who criticized the over-exploitation of whales. Indeed, for most of the history of Antarctica's whaling industry, it was probably the case that more scientists were deployed in the interests of killing whales (such as improving harpoons) than in researching their life-cycles and habitats.

Third, public opinion played the key role in convincing governments to take more notice of the scientific warnings. In Australia's case, for example, the pledge to stop whaling in Australian waters, which was honoured in 1980, became a major election pledge of a conservative government (Malcolm Fraser).

Fourth, whalers are tenacious. They may suffer reversals and have to endure moratoria. But if there are whales still alive, they will still try to find ways of killing them – even if this is done for 'scientific' purposes.

Finally, Antarctica has provided problems for policing regulations. It is far easier for a government to police its own waters to stop or restrict whaling, than to police an open area such as the Southern Ocean; indeed, it is far more difficult even to monitor what is going on in the Southern Ocean. Any attempt to devise a new regime for Antarctica has, then, to pay attention to the verification of international agreements.

Antarctic seals

The recent history of sealing contains a similar tale of governmental obstinacy. Under the umbrella of the Antarctic Treaty, the Consultative Parties agreed on a number of measures to protect the flora and fauna of Antarctica. The Treaty does not, as noted, provide for the protection of marine living resources in the waters off the Antarctic continent.

Sealing was another Antarctic industry. Southern elephant seals and southern fur seals were heavily exploited in the 19th and early 20th centuries and the populations of both were seriously depleted. But as fashions changed so the market changed and the exploitation ceased several decades ago. Populations have now recovered almost to their estimated pre-exploitation level. In 1964, however, Norway raised the possibility of commercial sealing. The Consultative Parties began work on a treaty to stop commercial sealing.

In 1972, the Consultative Parties held a special diplomatic conference in London at which they adopted a Convention for the Conservation of Antarctic Seals. The Convention – which came into effect in 1978 – required contracting parties (that is, governments which ratified it) to control the sealing activities of their nationals so as to ensure that any sealing was done according to regulations over what type of seals could be caught, which could not be caught at all, at what time each year sealing could be conducted, in what areas, and by what methods, with reports being submitted to the Scientific Committee on Antarctic Research of the International Council of Scientific Unions (SCAR).

The creation of that treaty had three advantages over the continuing campaign to stop whaling. Only one nation raised the possibility of sealing, rather than several. Second, Norway lacked the political and economic muscle of Japan and the USSR. Third, it was easier to stop an industry from starting – rather than curtailing a flourishing industry.

Sealing, since 1972, has therefore not been a major matter. A problem arose, however, in 1987 when it was reported that Soviet vessels had captured almost 5,000 seals. In August 1987, New Zealand threatened to ban Soviet polar vessels from its ports as a retaliation against this sudden upsurge in sealing. As is often the case, opinions differed as to what was happening. Tim Gentle, a New Zealand spokesperson for Greenpeace, said, 'All the seals which were taken were apparently used. The meat was taken back to the Soviet Union and all the carcasses were used within the Soviet Union.' Meanwhile a Soviet embassy spokesperson, Vladimir Ivanov, denied that any conservation agreements had been broken when two vessels went on sealing expeditions in December 1986 and January 1987. He said the Soviet Union had remained within the Convention, which stated that Antarctic seals could be killed or captured only to provide indispensable food for humans or dogs, for scientific research or for specimens for museums or other institutions. Overall, 5,010 crab-eater seals were killed. Thirty Ross and two elephant seals, both protected, were taken, along with a number of sea leopards and weddell seals. Ivanov said thousands of museums in the Soviet Union need seal specimens for display and research. Members of the Sealing Convention had been informed of the killings, and scientific data from the exercise would be made available in October. It is understood that nations belonging to the Antarctic Treaty group, including Australia, were also told.[6]

This incident prompted Australia to call for a review meeting of the Convention (which was in any event due in 1988 as provided by the Convention every five years). The UK supported this call. The meeting took place in September 1988 in London, ironically, at a time when seals were major news items as thousands were at the time dying in the North Sea from an unexplained virus or a form of industrial pollution. The meeting was not satisfactory. The USSR had not been as candid in its reporting as some governments (and environmental non-governmental organizations) would have liked. An amendment to strengthen the Convention was politically unacceptable as it could create a precedent for amending the Antarctic Treaty after 1991. Equally, there was no clarification of how many seals have to be caught each year before an activity moves from being 'research' to a 'commercial' venture. The USSR claimed again that its work was not of a commercial nature. Other nations (with Japan remaining silent) indicated that they were unlikely to undertake commercial Antarctic sealing in the foreseeable future.

The recent experiences in Antarctic whaling and sealing contain three general lessons. First, any treaty (or any other agreement) that permits 'research' catches runs the risk of being abused; particularly when implementation of the treaty is left in the hands of national governments (as distinct from, say, an international organization issuing permits for research catches). Second, scientists employed by governments (as distinct from international organizations) may give advice that is tempered by national political considerations. There is much which scientists not on government payrolls still do not know about Antarctica, and so the problem of the lack of scientific agreement cannot be minimized even among scientists trying to be politically impartial. But governmental scientists may not always be as impartial as could be hoped for.

Third, the international community (such as governments and non-governmental organizations) have few ways of coercing a government to stop behaving in a certain way. All these points will recur in the context of conserving the Antarctic marine living resources.

Agreed measures

In 1964 the Consultative Parties adopted the Agreed Measures for the Conservation of Antarctic Fauna and Flora. The subjects of protection are not 'resources' in the same class as whales and seals. But the document should be noted in this chapter as an example of how the Consultative Parties have tried to protect Antarctica's ecosystem.

According to the Agreed Measures the Antarctic Treaty area is to be considered a 'Special Conservation Area'. Participating governments are to prohibit within this area the killing, wounding, capturing or molesting of any native mammal or native bird – unless a permit to do so has been issued.

Permits may be issued only to provide specimens for scientific study, for educational purposes or for cultural uses or to provide essential food for humans and dogs while in the Treaty area.

The Agreed Measures remain important. First, they ban hunting as an entertainment, commercial undertaking or a recreation. Second, they require Antarctic residents to guard their dogs against killing or wounding wildlife. They emphasize that Antarctic activities should take place in the context of respecting Antarctica's wildlife.

The weakness of the Agreed Measures is, of course, their reliance upon individual governments honouring their obligations. Permits are issued by each government; the issuing is not subject to consultation with other Consultative Parties.

Antarctic living resources

Following a reduction in Antarctic whaling, commercial fishing emerged as an important new activity in Antarctic waters. There were, however, fears that fishing, particularly of krill, which occupies a central position in the marine food chain, might pose a threat to the entire marine ecosystem if it were not properly regulated. The Consultative Parties decided to develop a broad management regime for Antarctic living marine resources. In 1980 work was completed on the Convention on the Conservation of Antarctic Marine Living Resources (CCAMLR).

This chapter sets out the background to CCAMLR. In this section Antarctica's living resources, especially krill, are examined. CCAMLR is then examined with an assessment of how well it is working. The weaknesses of CCAMLR were not a good omen for the proposed treaty on mineral resources. The nation most responsible for the Convention's lack of progress is the USSR, one of the world's more important fishing nations. *Perestroika* and *Glasnost* are taking a long time to reach Antarctica.

Most of the world's population lives north of the Equator. Given the perishable nature of the catch, most of the world's fishing throughout history has been done well away from Antarctica. Only in recent decades has fishing in Antarctica become a viable industry – due both to improvements in fishing technology and in the decline of fish stocks and open fishing areas in the traditional northern regions.

Before dealing with Antarctica, however, it is useful to have an overview of world fisheries production. The International Centre for Ocean Development is a Canadian Crown Corporation which initiates, encourages and assists programmes in developing countries for the development and management of their ocean resources. The Centre has published a survey of world fisheries production as at the mid-1980s.[7]

Total world catch in 1983 was 76,471,000 metric tonnes; this total apparently includes freshwater fisheries and fish farming but by far the greatest part is marine wild fish. Fish supply about 13 per cent of the world's animal protein – about twice as much as eggs and three times as much as poultry. While fewer than 60 million people consume more than two-thirds of the world's meat and dairy products, fish are the primary source of animal protein for more than half of the world's people. In such countries as India, fish are a crucial part of the diet. Even in the meat-eating countries fish are important because of the extensive use of fish-meal as food for livestock. Only about 35 per cent of the total world fish harvest is actually consumed directly by humans. More than 30 per cent is used for purposes that include oil, fertilizer and livestock meal.

Fishing, then, is big business. All the major fishing nations with advanced technology for distant fishing have large fleets in the Southern Ocean: East and West Germany (now united in one Germany), Japan and the USSR.

Antarctic ecosystems are uniquely dependent on the marine environment. The food chains begin in the ice pack region of the surrounding Southern Ocean where the nutrients and high levels of carbon dioxide and oxygen create some of the most productive waters in the world. A rich crop of phytoplankton is fed upon by an abundance of krill and other zooplankton. Krill stocks are a critical link in the marine food chain and are the primary food source for many Antarctic fish, birds and marine mammals.

The main target of the Antarctic fishing industry now is krill (*Euphausia superba*), which looks like a shrimp. Antarctic krill occurs only in the Southern Ocean: the 36 million square kilometres of sea that surround Antarctica from the shore to as far north as the Antarctic Convergence. (At the Antarctic Convergence, which is located in the region of 55–60 degrees South, cold Antarctic water sinks beneath sub-Antarctic water, with the result that the Convergence forms a natural barrier for plankton.) Krill probably occurs everywhere in the Southern Ocean, but its abundance varies very greatly. Within the regions where it is common, the krill forms into large numbers of extremely concentrated swarms. Each may contain as much as several kilograms of krill per cubic metre. The swarms vary in size from a few metres to several hundred metres across, and it is this swarm-forming habit that makes it relatively easy to trawl for krill.

Krill swarms appear to occur most where they have the most plentiful supplies of their phytoplankton food. In its turn, this phytoplankton grows in the waters most richly endowed with nutrients. The minute one-celled plants that make up the phytoplankton contain chlorophyll, which fixes energy from the sun for use by the cell. The Southern Ocean is the world's most prolific producer of plankton, and parts of it are very rich in nutrients.

Soviet and Japanese fishing vessels are the most active in catching krill. They make it into a highly nutritious fish paste. The Japanese also convert krill into a protein juice and also make a form of mince out of it. Krill is an acquired taste; but even the British are acquiring it – potted krill tails have been sold at Harrods.

But a krill-fishing industry is no automatic bonanza. Krill perishes quickly and so needs to be processed on board the fishing vessels. Krill fishing is an expensive, capital-intensive industry which will require 'factory' vessels designed specifically for krill fishing. This will not be a route to ending hunger in Third World nations, since few of them could afford such an industry – assuming that their people could acquire a taste for krill.

Although the exact extent of the resource is unknown, scientists estimate the stock of krill at some 150 million tonnes. By comparison, the total amount of fish caught each year around the world is about 75 million tonnes. Krill could, then, be the most abundant source of animal protein the world has yet to tap. Less than 500,000 tonnes are caught each year at present. Research is continuing into the biology and life-cycle of krill. In a world where catches from traditional fishing grounds have been in decline and many grounds closed-off to foreign operators, the interest in exploiting this resource is understandable. In turn, this interest in the exploitation of krill has raised fears that over-harvesting could permanently damage the fragile Antarctic ecosystem.

Fears about the effects of a depleted krill stock, together with longstanding concerns about other species such as whales, led Antarctic Treaty signatories, scientists and conservationists to the belief that a formalized organization was needed for the protection of the Antarctic marine ecosystem. Humans are not the only creatures interested in krill. There is a need to balance potential human consumption against the current needs of whales, seals and penguins, which depend upon krill as their major food source.

Conference on the Conservation of Antarctic Marine Living Resources

The Conference concluded in Canberra, on 20 May 1980, with the signature of the Final Act. Those participating in the Conference were the then thirteen Antarctic Treaty Consultative Parties, as well as West and East Germany. The Commission of the European Community and a number of inter-governmental and non-governmental organizations were represented by observer delegates.

The Conference ended three years of negotiations on the marine living resources issue. Since the negotiations got underway at the Ninth Antarctic Treaty Consultative Meeting in London in 1977, there were seven separate meetings and consultations on the draft Convention. The key articles in the Convention were the subject of a 'gentlemen's agreement' reached in Washington in September 1978. Stumbling blocks had included French insistence on a separate understanding spelling out the manner in which France would exercise its sovereign rights around Kerguelen and Crozet islands under the new regime. The other stumbling block was the manner in which the European Community Commission should participate in the Convention. The USSR opposed the role the European Community Commission proposed for itself in both the Convention and the Conference. It was eventually settled that the European Community Commission would not be an original signatory to the Convention, but that there would be an understanding that it should be assured of a place on the proposed Commission. With European politicking settled, the Conference could get on with its real task of finalizing the treaty.

The Convention on the Conservation of Antarctic Marine Living Resources contains 33 Articles and an Annex for an Arbitral Tribunal for disputes among two or more Contracting Parties.

Article I defines, in a novel way, the geographical area of application of the Convention. The Antarctic Treaty covers the area south of 60 degrees South Latitude. This neat, straight-line approach made sense because it was primarily concerned with the Antarctic land mass. This Convention, however, includes all the water within the Treaty area but goes beyond it to the Antarctic Convergence – whose latitude varies. Thus, the geographical area of application is not based on straight lines so beloved by lawyers but on the curvy outer limits of the Antarctic marine ecosystem. This ecosystem is defined as 'the complex of relationships of Antarctic marine living resources with each other and with their physical environment'. According to Article II, the term ' "conservation" includes rational use'. The Article continues:

> 3. Any harvesting and associated activities in the area to which this Convention applies shall be conducted in accordance with the provisions of this Convention and with the following principles of conservation:
>   a. prevention of decrease in the size of any harvested population to levels below those which ensure its stable recruitment. For this purpose its size should not be allowed to fall below a level close to that which ensures the greatest net annual increment;
>   b. maintenance of the ecological relationships between harvested, dependent and related populations of Antarctic marine living resources and the restoration of depleted populations to the levels defined in sub-paragraph (a) above; and
>   c. prevention of changes or minimization of the risk of changes in the marine ecosystem which are not potentially reversible over two or three decades, taking into account the state of available knowledge of the direct and indirect impact of harvesting, the effect of the introduction of alien species, the effects of associated activities on the marine ecosystem and of the effects of the environmental changes, with the aim of making possible the sustained conservation of Antarctic marine living resources.

Under Article III, the Contracting Parties, whether or not they are Parties to the 1959 Antarctic Treaty, 'agree that they will not engage in any activities in the Antarctic Treaty area contrary to the principles and purposes of the Treaty . . .' Article IV reaffirms the 1959 Treaty provision on, in effect, freezing territorial claims.

Under Article V, nations which are parties to the Convention but are not parties to the 1959 Treaty 'acknowledge the special obligations and responsibilities of the Antarctic Treaty Consultative Parties for the protection and preservation of the environment of the Antarctic Treaty area'. Article VI reaffirms the rights and obligations of Contracting Parties under the International Convention for the Regulation of Whaling and the Convention for the Conservation of Antarctic Seals.

CCAMLR creates two major organs to assist in its application: the Commission for the Conservation of Antarctic Marine Living Resources (Articles VII–XIII) and the Scientific Committee for the Conservation of Antarctic Marine Living Resources (Articles XIV–XXIII). The Commission, with its Headquarters at Hobart, Tasmania, has delegates from all the Contracting Parties, and 'regional economic integration organizations' (such as the European Commission). The Commission itself is, in effect, a political organ. The Commission's secretariat is an international civil service. Scientific advice is provided by the Scientific Committee with members drawn from all the members of the Commission.

Also to assist CCAMLR's implementation, Article XXIV provides for a system of observation and inspection. Article XXV and the Annex deal with the resolution of disputes between Contracting Parties. Articles XXVI–XXXIII deal with the standard procedural provisions found in all multilateral treaties.

CCAMLR – an assessment

The Convention has, at first sight, some remarkable features. It has a geographical area of application which accords with the interests of scientists (the Antarctic Convergence) rather than lawyers (who prefer neat lines on maps even if they do not quite fit geographical features such as 60 degrees South Latitude). Second, the ecosystem approach was also an innovation. It is based on the concept of viewing each resource in its environmental context, rather than on its own. This stresses the interrelated and interdependent nature of the web of life. Third, it has reaffirmed the basic ideas of the 1959 Treaty on not permitting further territorial claims or forcing current claimants to forego their claims.

Fourth, it has established a new international organization to assist its implementation. The Convention provides for a permanent secretariat in a way that the 1959 Treaty does not. The Scientific Committee's status, on paper anyway, demonstrates the significance accorded by the Contracting Parties to the value of research.

Finally, the Convention demonstrated to the international community that the Antarctic Treaty Consultative Parties could be trusted to govern Antarctic affairs responsibly. They did the basic work for the Conference and have had their special status enshrined in CCAMLR. The 1959 Treaty is, then, a firm foundation upon which other treaties could be constructed. The next issue would be minerals. After that treaty, according to this viewpoint, separate treaties could be devised for (say) tourism and icebergs. Each would build upon and reaffirm the value of the 1959 Treaty.

On the other hand, from a conservationist point of view, CCAMLR has some worrying features. First, its very title suggests that the marine life is to be seen as 'living resources' – the word 'resources' implies a supply of goods to be used for commercial purposes. Second, this worry is taken up explicitly in Article II(2) whereby ' "conservation" includes rational use'. Thus CCAMLR talks about conservation but it is, in fact, a regime for exploitation. The exploitation is, of course, to be done 'rationally' – but it is still exploitation.

Third, soon after the Convention was finalized, Australian international lawyer F.M. Auburn sounded a warning voice. Although CCAMLR's conservation principles read well, he predicted that there would be serious obstacles to any attempt to introduce specific measures binding nations which wish to conduct commercial activities:

> There is no specific funding mechanism for scientific research on krill stocks, and the Convention does not prescribe catch levels or any other form of regulation. Krill trawling countries can either veto conservation measures or individually refuse to accept them. The Scientific Committee is essentially under the direction of the Commission, which represents the political interests of States. Impact assessment in the municipal law sense is not provided for under the regime, and even if the Scientific Committee were to reach specific conclusions on, say, desirable catch levels, such findings would in no way bind the Commission.[8]

Fourth, there is no international inspection to verify the information on catches. In this respect, then, the Convention is weaker than the Antarctic Treaty. If fishing restrictions are violated, then the flag state is responsible for prosecutions and sanctions; evidently few – if any – prosecutions have taken place.

Fifth, there is no recognition that krill could be a resource as part of the 'common heritage of humankind'. There is no obligation to share the catch with protein-needy Third World nations.

Finally, the Convention augments the special status of the Antarctic Consultative Parties. For example, Article V of CCAMLR requires parties to the Convention to acknowledge the special obligations and responsibilities of the Antarctic Treaty Consultative Parties for the protection and preservation of the environment of the Antarctic Treaty area. It also compels them to agree to observe, as and when appropriate, the Agreed Measures for the Conservation of Antarctic Fauna and Flora, and such other measures as have been recommended by the Antarctic Treaty Consultative Parties. This language particularly attempts to bind non-Antarctic Treaty nations to future actions of the Consultative Parties. The Consultative Parties also decided on criteria for future participation in the Convention that allowed themselves to be granted permanent voting rights. All other nations desiring to be voting members of the Commission established by the Convention have to pass active research and harvesting tests. Article XXIII of CCAMLR requires the Commission and its Scientific Committee to co-operate with the

Consultative Parties 'on matters falling within the competence of the Consultative Parties'. But the competence is not indicated, leaving it open for intervention by the Consultative Parties on any pretext.

CCAMLR in action

CCAMLR came into force in April 1982. The following nations are Contracting Parties: Argentina; Australia; Belgium; Brazil; Chile; France; East Germany; West Germany; India; Japan; South Korea; New Zealand; Norway; Poland; South Africa; Spain; USSR; United Kingdom; USA. The European Community is also a Contracting Party.

CCAMLR's first seven years have justified the critical worries that some conservationists have had about its effectiveness. The problems predicted by Professor Auburn have all arisen in one form or another.

The first full meeting of the Commission was held in Hobart in 1982 and annual meetings have been held since. The first two meetings of the Commission were primarily concerned with legal, administrative and financial aspects of the establishment of the Commission as well as initial consideration by the Scientific Committee of the steps required to provide an adequate scientific basis for subsequent advice and recommendations. A major problem facing the Commission is the lack of data and research concerning many aspects of the Antarctic. More information is necessary concerning both the extent of the resources to be managed and conserved as well as the relationship of each particular species with the ecosystem as a whole. The problem is exacerbated by the hostile conditions experienced in Antarctica for most of the year.

The 1984 meeting made the first tentative conservation steps by adopting binding measures to ban all fishing within 12 nautical miles of the island of South Georgia and limits on net mesh sizes to protect juveniles of Antarctic fin fish. Member nations were also requested to ensure that no directed fishing for Antarctic Cod (*Notothenia rossii*), a species of fin fish whose stocks are believed to be severely depleted, took place in the South Georgia area. The 1985 meeting of the Commission made further limited progress towards the conservation of Antarctic Cod by prohibiting any directed fishing for Antarctic Cod in the South Georgia area. Members were also requested to cease fishing in the South Georgia area if the total by-catch (that is, fishing operations directed at other species) of Antarctic Cod exceeds 300 tonnes, and to refrain from directed fishing for this species in the Peninsula, South Orkneys and Kerguelen areas.

This rate of progress (if that is the right word) caused annoyance among governments (such as Australia's), concerned about conserving Antarctic marine life, and among environmental non-governmental organizations. An anonymous writer (who evidently knew what was happening behind the closed doors) described the 1985 deadlock in an Australian conservation magazine:

> It had already become apparent at last year's [Commission] meeting that overfishing in the South Georgia area, principally by the Soviet Union, had led to a serious decline in stocks of several species of finfish. One species *Notothenia rossii* (Antarctic cod) had been reduced to virtual commercial extinction. Yet, despite this evidence, the meeting could only adopt control measures already voluntarily followed by the Soviet Union and therefore not to be expected to produce significant change.
>
> Article 11 of the CCAMLR agreement explicitly provides the procedure to be followed in the instance of stock decline. The Commission's inability to take remedial action was not encouraging for the workability of the Convention.
>
> At this year's [1985] meeting, many members supported initiatives to introduce more effective fisheries control measures. However, the Commission was hamstrung by its consensus decision-making process when USSR, supported by Poland and East Germany, successfully blocked moves to follow Scientific Committee advice and close the South Georgia and Kerguelen fisheries. . . .
>
> Many delegations were extremely concerned about the fate of both the fishery and the future direction of the Convention itself. Australia issued a document which eloquently expressed their frustrations at the Commission's lack of progress over the past four years. Calling for a much more stringent adherence to the basic principles of the Conventions, the Australians should have received substantial support from the non-fishing nations. Astonishingly, only New Zealand spoke in strong support of the Australian position. Other nations which had spoken out for sound conservation measures during the Scientific Committee, like the United Kingdom and the USA, failed to speak out at the critical time and thus missed their opportunity to salvage some credibility for their commitment to this unique international fisheries agreement.[9]

The fifth Commission meeting in 1986 was also disappointing from a conservationist viewpoint;.

> It would appear from the results of the fifth CCAMLR meeting that yet again the Commission has failed to take action to implement Article 11 of the Convention. The Scientific Committee advice has been ignored. The fish at South Georgia have been

> severely depleted. At the end of the meeting, at the CCAMLR press conference, the Commission Chairperson stated that directed fishing of Antarctic Cod, *Notothenia rossii*, is to be banned from major fishing grounds in the South Atlantic Ocean.
>
> Unfortunately, this will not have the desired result of completely stopping the catches of *Notothenia rossii* in this area – the only sure way to do that would be to go for the closure of the area. Incidental catch will still take a toll of this species.
>
> The failure of the CCAMLR has real and severe implications for the rest of the Antarctic Treaty system. If the system fails on a relatively simple matter like the control of marine resources, how then can it be expected to cope with the pressures that a large scale mining operation might impose on it? The Scientific Committee of CCAMLR has become highly politicized over marine resource issues; is it not then reasonable to assume that the mineral regime's Advisory Committee will be equally politicised?[10]

The sixth meeting, in 1987, was also disappointing. A note of weariness appears in the official report. CCAMLR itself seemed destined to become an endangered species. 'Several delegations expressed concern that members' budget contributions were not reaching the Secretariat by the 31 May deadline each year'.[11] Some delegations even suggested reducing the budget – with cuts falling in the area of scientific research.[12] The Commission's total budget is, incidentally, about US$1 million. If money is an indication of how cursorily the Contracting Parties view CCAMLR, then obviously CCAMLR has a low priority.

The 1988 meeting, from a conservationist viewpoint, was a bit of an improvement (which is not necessarily saying very much). The editorial in the ad hoc environmental newsletter produced for the meeting stated, 'for the first time in seven years, *ECO* finds itself in a position to extend a sincere message of appreciation' to the CCAMLR meeting 'for the work being done in the Scientific Committee as well as the Commission'.[13] Moving at glacial speed, the meeting made some progress on the creation of an inspection and observation system. It also permitted the non-governmental Antarctic and Southern Oceans Coalition (ASOC) to attend the plenary sessions as an observer. ASOC represents 172 environment non-governmental organizations in 35 countries and has lobbied outside the closed doors of all CCAMLR meetings. It has been concerned by many of CCAMLR's decisions which run contrary to scientific advice or are made in the absence of necessary information. ASOC has called for a moratorium on all Antarctic fishing until all species have recovered sufficiently to support a sustainable catch. It was also alarmed at the failure of CCAMLR to regulate krill fishing.

The eighth meeting in November 1989, alas, fell back to the old pattern. Lyn Goldsworthy attended on behalf of ASOC. In regard to krill she reported that the total catch of krill in the 1988–89 season was 395,470 tonnes compared to 370,663 tonnes in 1987–88 and 376,456 tonnes in 1986–87. The USSR and Japan remained the major harvesting nations, catching 300,000 tonnes and 80,000 tonnes respectively. Chile, Republic of Korea and Poland each caught less than 10,000 tonnes. Japan indicated that it did not expect to increase its catches as there was little market interest in the product. In contrast, the USSR announced that it had solved its processing problems and was likely to expand its harvesting substantially over the coming years both in the amount caught and the areas in which they would be operating. ASOC had information that Finland was building several large, integrated vessels suitable for fishing in the Southern Ocean, for the USSR. These could dramatically increase the fishing capabilities of the Soviet fleet.

With regard to fishing, the USSR was also the leading nation in being obstructive. On fin fish, for example, the USSR accepted that scientific data remained unclear as to the total stock available – but then claimed given that there was a lack of data no management procedures should be enacted. Ms Goldsworthy reported, 'It is obvious that the USSR is not seeking to fulfil the objectives of the Convention. The USSR position is clearly not based on a rational-use long-term analysis but on a short-term "rape and plunder" attitude.'[14]

Ms Goldsworthy ended her report:

> This meeting shows once again that the potential of the 'ecosystem as a whole' conservation requirement of CCAMLR is not being met. Fishing states are given the power to block rational conservation measures and other states lack the diplomatic willpower to counter the intransigence of the fishing states. NGOs should be calling for a major international review of the functioning of CCAMLR, and the adoption of new rules, based on the precautionary principle, that would establish the entire fishery on a truly scientific basis. This has been the ASOC position since CCAMLR was negotiated.[15]

To conclude, CCAMLR's record does not inspire confidence in the future management of Antarctica. Some governments, such as the UK's, Australia's and New Zealand's, are obviously trying to restrict the over-exploitation of Antarctica's living resources. But they can do little in the face of obstinate nations. The omens for a parallel mineral regime are not good.

Notes

1. Creina Bond and Roy Siegfried *Antarctica: No Single Country. No Single Sea* London, Hamlyn, 1979, p. 108.

2. Simon Lyster *International Wildlife Law: An Analysis of International Treaties Concerned with the Conservation of Wildlife*, Cambridge, Grotius, 1985, pp. 17-38.

3. Keith D. Suter 'The International Politics of Saving the Whale' *Australian Outlook*, Canberra, December 1981, pp. 283-4.

4. 'Moratorium Fails to Stop Killing of Whales' *The Australian* 30 June 1990, p. 5.

5. 'Japanese Decision on Driftnetting Welcomed' *The Australian*, 18 July 1990, p. 1.

6. 'Seal Row May Shut Ports', *Canberra Times*, 4 August 1987.

7. 'World Fisheries Production' *Maritime Studies*, Canberra, March 1990, p. 26.

8. F.M. Auburn, *Antarctic Law and Politics*, Canberra, Croom-Helm, 1982, p. 289.

9. From 'A Special Contributor', 'Soviet Over-fishing Hits Antarctic Cod', *Australian Conservation Foundation Newsletter*, Melbourne, December 1985, p. 10.

10. Margaret Moore, 'CCAMLR's Failure to Jeopardize Antarctic Treaty System?', *ACF Newsletter*, November 1986, p. 12.

11. Commission for the Conservation of Antarctic Marine Living Resources, *Report of the Sixth Meeting of the Commission*, Hobart, 1987, p. 4.

12. Ibid., pp. 4-5.

13. '*ECO* Applauds Progress', *ECO* (Friends of the Earth/Greenpeace International/Antarctic and Southern Oceans Coalition) Vol. XLI, No. 3, October 1988, p. 1.

14. Lyn Goldsworthy 'Report on the Eighth Meeting of the Commission for the Conservation of Antarctic Marine Living Resources' *Antarctic Society of Australia Newsletter*, Sydney, March 1990, p. 5.

15. Ibid., p. 8.

4. Antarctic Minerals

'Don't sign – don't mine!'

After a meeting of the Antarctic Treaty Consultative Parties in Wellington, New Zealand, in 1972, a European delegate is reputed to have said: 'This treaty will last till a big mineral discovery is made – then it will be every man for himself.'[1]

Eighteen years on, that prediction still seems valid. Minerals are known to exist in Antarctica but no big mineral discovery has been made. With the treaty on living resources completed, the Consultative Parties in 1982 began work on a similar treaty for mineral resources; the work was completed in 1988. At the signing ceremony in Wellington in November 1988, diplomats were greeted by chants from demonstrators: 'Don't sign – don't mine'. This comparatively small demonstration was reported in the mass media in several nations – in itself evidence of the media's increasing awareness of Antarctic controversies.

There is now confusion as to the future of any minerals regime. On 22 May 1989 the Australian Government announced that it would not sign the minerals treaty. All claimant nations – Argentina, Australia, Chile, France, New Zealand, Norway and the UK – had to ratify the Treaty before it could enter into force. Australia was the first one of the seven to refuse to do so.

This refusal began a trend. France very quickly followed Australia's example. They began a joint campaign for a legal framework of environmental safeguards to prevent mining. As at July 1990, the US and UK oppose the Australian–French plan but other nations of varying importance to the Antarctic Treaty System are sympathetic. Chile and New Zealand will not sign the new Treaty and they have pledged to work towards a compromise. India, Italy, Belgium, Greece, Austria, Hungary, Romania and Bulgaria also support the mining ban. The US Administration itself still supports the new Treaty but two congressional bills working their way through committee call on the US to abandon the Treaty, while other legislative initiatives gaining support mirror the Australian–French position. The British Opposition supports Australia's approach and the USSR may be moving towards the protected nature reserve idea. Japan has said nothing.

However, even if the Treaty is dead, there still remains a desire among some nations to have some mineral development. The environment movement, which has scored a remarkable achievement in ending the Treaty, still has much work to do.

An Antarctic El Dorado?

Antarctica used to be joined to Africa, South America, Australia, India and New Zealand. This was the supercontinent Gondwanaland. Two hundred million years ago, Gondwanaland broke up with most of the fragments each drifting its own way north. Gondwanaland's core – Antarctica – drifted south. The new – and much colder – climate resulted in the extinction of Antarctica's life. In recent years, scientists have found fossilized ferns, freshwater fishes, amphibians and reptiles in Antarctica. The geographical separation, and cold, dry climate that developed, prevented many groups of plants and animals on other continents from invading and colonizing Antarctica: for example, trees, flowering plants, land-based reptiles and mammals. Antarctica is the only continent where humankind is not indigenous and where civilizations have never developed.

Antarctica's 'wealth' may be viewed in various ways. One way, which is often obscured by the increasing speculation over krill and minerals, is Antarctica's unique value to science:

> A major find this [1985–86] season came in the Transantarctic Mountains, the 1,900-mile-long range that cuts across Antarctica like the spiny plates of a stegosaurus. There, scattered around a small, snow-free valley, researchers found petrified remains of a 220 million-year-old swamp. The flora were so faithfully preserved that the scientists identified the impressions of leaves, seeds and branches from a tree like the Norfolk Island pine and even parasitic fungi. In a single day they loaded the helicopter with two tons of samples that, according to team leader Thomas N Taylor of Ohio State University, were 'worth more than gold'. Many of the fossil finds in the Transantarctics seem to have no business on a continent only 2 per cent of which is ice-free. Taylor's swamp fossils show that Antarctica was once in a temperate clime; they can be matched to fossils on other continents to get a more precise picture of how land masses used to fit together.[2]

Second, as the previous chapter has indicated, the Southern Ocean has been a consistent source of wealth for a century: whales, seals, fish and krill. Marine activities have been difficult and dangerous but not necessarily much beyond what would have confronted mariners conducting similar operations in the northern hemisphere.

Third, there is increased speculation about mineral wealth in Antarctica. The earliest explorers on the continent itself discovered minerals, and Antarctic coal was used by those explorers. This chapter looks at the current interest in the mineral wealth of the rock itself and land under the surrounding ice-shelf.

The fourth and fifth sources of potential wealth are tourism and the towing of icebergs to areas lacking freshwater. These issues are examined in Chapter 10.

Antarctic minerals

All the other parts of Gondwanaland are rich in minerals. Does Antarctica also contain minerals? In theory: yes. But the presence of minerals alone is not a satisfactory indication of wealth. There has to be a world market for the minerals, an economical method of extraction and transportation, and a firm legal basis for operating in a territory 'claimed' by a government (even if most of the world's governments do not recognize that claim) – and all this has to be done in the context of not eroding Antarctica's value for scientific research.

> The mountains of the Antarctic Peninsula were once continuous with the Andes, which means that the former may well have lead and copper deposits similar to those in Peru and Chile.
>
> The Ross Sea continental shelf was once adjacent to the Bass Strait between Tasmania and Australia, where oilfields have already been discovered. Oil has also been found off New Zealand and Argentina, which were adjacent to Antarctica in Gondwanaland.
>
> Structural similarities between the East Antarctic shield and parts of Australia and South Africa suggests the possible occurrence of uranium.[3]

Interest in Antarctica's potential mineral wealth has been slow in emerging. Until recently this was a matter of interest only to scientists and arose in the context of other scientific investigations, such as the speculation over Gondwanaland's existence.

Additionally, there has been no economic incentive to explore for minerals. Whatever minerals Antarctica did contain could be obtained far more cheaply elsewhere around the globe – not least from some of the Antarctic Treaty Consultative Parties themselves. These nations, such as the US and Australia, obviously wanted to sell their minerals before exploration began in Antarctica.

There have also been practical problems of extraction – Antarctic mining was beyond the technological capacity of all nations and corporations. Antarctica is largely covered by an ice sheet which is sliding towards the sea slowly. Mineral activity would be possible in only very limited areas of onshore Antarctica. Only 2 per cent of the continent consists of exposed rock, the remainder being covered by ice, on average 2.5 kilometres thick. So far, with the exception of low grade coal and iron ore, no mineral deposits large enough to be even potentially economic have been found. There may be greater potential for offshore hydrocarbon exploitation, although, again, the extent of any such resources in Antarctica is unknown. Research has shown the presence of sedimentary basins in the Ross and Weddell seas and in the Prydz Bay area off the Australian Antarctic Territory. Shallow drilling has confirmed the presence of some hydrocarbons in the Ross Sea continental shelf. Considerable exploration would be required before it could be determined whether hydrocarbons were present in commercial quantities. Even if found, there would remain some doubt as to whether exploitation of hydrocarbons in Antarctica would be technically possible, safe and commercially viable; for example, the moving ice sheet would snap any drill.

Besides, even if minerals did exist in Antarctica, there is no guarantee that they would be in sufficiently large deposits to make even the hint of prospecting worthwhile. To be economically attractive, minerals in the Antarctic continent would need to be large deposits of high concentration and value, with good access to transportation.

The situation began to change somewhat in the early 1970s. The 1972 UN Conference on the Human Environment publicized both the adverse changes to the environment and the possibility that the globe could be running out of easily accessible non-renewable resources. The US research ship *Glomar Challenger* in 1973 discovered traces of hydrocarbons in the continental shelf off the Ross Dependency. In October 1973 the Organization of Petroleum Exporting Companies (OPEC) increased the price of oil. OPEC was formed in the 1960s to safeguard oil prices (which in some years actually declined in comparison with goods brought from Western nations). The strength of this cartel was increased during the 1973 Arab–Israeli conflict, when, for a time, the Arab OPEC members cut oil sales to Western nations. This isolated Israel from its Western supporters and increased the price of oil. A further wave of increases began in 1978 when Iran's exports were reduced during the chaos of the Shah's last months in power. Developed Western countries realized how vulnerable their reliance on Arab oil supplies made them, and so they found it politically expedient to locate alternative oil supplies. Quite suddenly, oil fields which previously were of marginal significance became viable. Thanks to North Sea oil, for example, the UK became virtually self-sufficient in oil by 1980.

Meanwhile, extractive technology has improved. North Sea oil production – which was at one time at the outer limit of this technology – proved, providing the world market price was right, that oil drilling even in the North Sea was possible and profitable. Improvements were also made in the extraction and transportation of natural gas. Natural gas now provides the world with an amount of energy roughly equal to half that of all the oil produced. Natural gas used to be transported by pipeline and only for short distances; with the exception of Canadian gas going to the US, the gas was consumed in the nation of extraction. Natural gas is now chilled into a liquified form and shipped from one nation to another (Australia's liquified natural gas (LNG) goes to Japan).

Extractive technology is certainly improving. But Antarctica still presents a challenge far greater than that confronted anywhere else on the globe. Barbara Mitchell has recently warned that Arctic extractive technology cannot be easily redeployed in Antarctica:

> One of the biggest problems is the icebergs that slough off Antarctica's ice shelves – which in winter extend about 1,000 miles past the shore. The icebergs that break off these shelves are much larger than their glacially formed counterparts in the Arctic. When an iceberg is bearing down on an oil-drilling rig in the Arctic, crew members often send out a tug-boat to tow it away. If the iceberg is larger than 100 cubic metres, they disconnect the rig from the well and leave quickly, later returning to reconnect. But about a third of Antarctica's waters have so many large icebergs – many of them a cubic kilometer – that oil-drilling operations would have to flee often.
>
> Furthermore, the deep water surrounding much of Antarctica would make it too costly to install the platforms that sit above production wells. Platforms are used to control the flow of oil, and often to separate it from other constituents rising from the wells. The deeper the water, the larger a platform must be to remain steady. In the Arctic, most platforms sit less than 300 feet above the seafloor. The water off much of Antarctica runs much deeper.[4]

Even if the economic and technological problems were solved, there are still immense legal problems. If a corporation did exploit the mineral wealth, to whom should the royalties be paid? How much should be paid? What safeguards should be introduced to protect the continent's environment? Since there is international co-operation in scientific research, should a corporation's operations be open to the same scrutiny? The questions seem endless. No corporation would even contemplate an investment of millions of pounds knowing that its work was resting on a legal quagmire.

Finally, the economic, technological and legal problems have had to be viewed in the context of safeguarding the continent for continued scientific research. Every stage of a mineral (including oil and gas) extraction project would contain threats to Antarctica's environment and would make scientific research all the more difficult.

The most obvious potential source of mineral wealth is oil. The greatest area of environmental concern is the possible effect of a major oil spill, which might result from tankers or supply ships being damaged by ice and subsequent spillage of fuel; a mishap to the drill ship or platform; fracture of the production pipeline; leakage from storage tanks; blowouts during drilling. The *Torrey Canyon* disaster of 1967 was an 'early warning' of what could happen.[5] The *Torrey Canyon* was a giant oil tanker used to transport oil from the fields of the Middle East to refineries in Europe or the United States. In March 1967, it ran aground off the coast of Cornwall and spilled its cargo of 36 million gallons of petroleum into the seas. The beaches were soaked, the holiday season was ruined, and thousands of seabirds became entrapped in oil and died, despite efforts to save them. In spite of a campaign to clean up the oil using detergents and napalm (themselves potential pollutants) extensive damage was done. The oil spread along the coasts of England and reached the beaches of France.

Even if the oil drilling went perfectly, then a 'perfect' operation would still have negative consequences. Even if the drilling was not done on the two per cent ice-free area of Antarctica, the workers would need to live on the area when not out on the oil rigs. They would be in competition for the land with the continent's flora and fauna. Ice-breaking ships near the oil rigs would create a deafening underwater noise, disrupting the migration patterns and breeding cycles of whales. There is also the potential impact on krill: they live only in the Southern Ocean, which is the world's most pollution-free marine environment – they have had, therefore, no incentive to cope with oil and so may have a low toleration to oil pollution.

The regulation of mining activities has not, therefore, attracted the same feeling of urgency as did living resources. The problem of the possible effects of minerals activity was first addressed by the Antarctic Treaty Consultative Parties in 1972. At subsequent meetings studies were made and principles agreed upon which could form the basis of a minerals regime. At the Ninth Consultative Meeting held in London in 1977 it was recommended that governments urge their nationals and other nations to refrain from exploration and exploitation of Antarctic minerals resources while making progress towards the adoption of a minerals treaty. This moratorium remains in place today pending the fate of the new minerals treaty – but the decision to end the new treaty must add to the confusion over the future of the moratorium.

Creating the Minerals Convention

The Consultative Parties' work on a treaty on minerals was undertaken via a series of Special Consultative Meetings. The July 1981 regular meeting of Consultative Parties created the mandate for this work. The regime should be based on the following principles: that the Consultative Parties should continue to play an active and responsible role in dealing with the question of Antarctic mineral resources; that the Antarctic Treaty must be maintained in its entirety; that the Consultative Parties, in dealing with the question of mineral resources in Antarctica, should not prejudice the interests of all humankind in Antarctica; that the provisions of Article IV of the Antarctic Treaty (relating to territorial claims) should not be affected by the regime; it should ensure that the principles embodied in Article IV are safeguarded in application to the area covered by the Antarctic Treaty.[6]

The negotiations began in Wellington, New Zealand, in June 1982 and were chaired by Chris Beeby, then Deputy Secretary of the New Zealand Ministry of Foreign Affairs. He remained the key chairperson and produced successive draft versions of the Treaty. There were twelve sessions of negotiations in all, culminating with the final one held in Wellington from 2 May to 2 June 1988. The participants were the Antarctic Treaty Consultative Parties. There were 14 Consultative Parties when the negotiations began in 1982; Argentina, Australia, Belgium, Chile, France, West Germany, Japan, New Zealand, Norway, Poland, South Africa, the USSR, the UK and the US. These countries were joined by Brazil, China, East Germany, India, Italy and Uruguay from the dates of the admission of these nations to consultative party status. In addition, it was decided in 1984 that nations that were not consultative parties should also be invited to attend the negotiations. As a result, Austria, Bulgaria, Cuba, Czechoslovakia, Denmark, Ecuador, Finland, Greece, Hungary, the Republic of Korea, Netherlands, Papua New Guinea, Peru, Romania, Spain and Sweden attended various sessions of the negotiations.

These negotiations were very complicated. That any document has been completed at all is a substantial achievement – but this does not mean that I endorse it. Indeed, the project has been flawed from the outset as (in my view) all commercial mining should be banned. The task, then, should have been one of finding a way of banning mining, rather than regulating it. But this does not suggest that the Consultative

Parties lacked imagination in tackling the negotiations – only that they put their imagination to the wrong use.

One theme in the negotiations has been the question that if mining and drilling were so far off in the future, why was there a need now to negotiate a regime? If such activities were several decades off, would it not be better to wait until more knowledge is gained of both the Antarctic ecosystem to be preserved and the new extractive technology to be employed? Just what has been motivating the Consultative Parties? This haste has created an air of suspicion among nations outside the Antarctic Treaty system that the Consultative Parties are closer to mining and drilling than they are revealing. Have they been negotiating a minerals treaty so as to keep the wealth between themselves or to keep out the Third World nations and the United Nations or, simply, because the Consultative Parties do not trust each other not to take advantage of the lack of a specific permanent ban on all mining?

In other words, the conspiracy in haste was not directed outside the Consultative Parties, but is simply a reflection of the lack of trust between themselves. For example, one or more claimant nations could permit mining as yet another way of trying to substantiate their territorial claims. Perhaps there is no fixed explanation, and each Consultative Party had its own motivations. What is quite clear, however, is that the haste has generated a suspicion which the Consultative Parties may well find self-defeating. After all, there is a crucial difference between living resources and mineral resources – a treaty is required for living resources since harvesting is already underway, but commercial mining and drilling are apparently not likely to take place for some decades.

Second, commercial mining and drilling have the potential of reopening the frozen issue of national claims to Antarctic territory. How could a regime be developed which would not somehow accord, by implication, a recognition to national claims by permitting some form of veto power to a nation which claimed a slice of Antarctic territory? Would nations like the US and USSR, which recognize no territorial claims, be drawn into recognizing claims by according some special status of claimant nations?

Third, an 'accommodation' had to be made between the Consultative Parties and other parties which were not permitted to be directly involved in the private negotiations of the Consultative Parties. Even further from the decision-making were nations which were not parties to the Antarctic Treaty, as well as international organizations such as the UN.

Fourth, how could the Antarctic environment be protected from the harmful effects of commercial mining and drilling? As stated earlier in this chapter, the more obvious efforts would come from oil spills. There could be a tanker collision, the rupture of a pipeline or a well-head by an iceberg, or a blowout of an oil well which could cause permanent damage to the fragile Antarctic environment. Because of the dependence of the entire Antarctic food chain on krill, an accident involving a release of oil into the environment might have far-reaching consequences. Any operation recovering oil from the Antarctic would need airstrips and shore facilities for personnel and for oil storage; these would necessitate extensive terrain changes. The areas of ice-free coastline that would be suitable for such facilities are extremely limited and already have been colonized by Antarctic wildlife. Interference with these coasts would put some colonies and the species which inhabit them in grave danger.

Less obvious but also of significance would be the distortion of funding for Antarctic research away from projects in which scientists are interested, into issues which would be of direct assistance to commercial activities. For example, Antarctic 'gold' for scientists is meteorites. Meteorites plunge to earth each day but are almost always burnt up in the atmosphere or, if they survive at all, fall unnoticed into the oceans. In 1969, Japanese glaciologists found nine meteorites in the Antarctic ice sheet. Meteorites have been falling for the past millions of years and have been trapped in the ice. A 'gold rush' got underway. Since 1969 about one half of all meteorites available for study in the world have come from Antarctica. This type of research has no commercial value and so would not attract much funding from governments – which would naturally be inclined to finance research projects with some possibility of financial return.

Fifth, by contrast, how could a regime be developed which would not deter commercial activities in Antarctica? Transnational corporations and government agencies could be deterred by such matters as excessive 'red tape', the deferral of decisions on political grounds (such as a Consultative Party not liking a particular corporation because of its activities elsewhere), a lack of guarantees that minerals discovered by a corporation could in fact be exploited by that corporation, an obligation on the part of corporations to share their extractive technology with Consultative Parties, and an obligation on the part of corporations to report fully and regularly on their prospecting and mining.

In other words, the treaty would need to offer inducements to transnational corporations for them to invest in Antarctic mining – while simultaneously establishing sufficient safeguards to protect the environment. Given the poor record of Western and Third World governments in being able to strike this balance within their own boundaries, it seemed most unlikely that they could achieve this task over Antarctic mining.

Additionally, international law is created by and for nations; transnational corporations are not the subjects of international law – how could a treaty be devised that would give rights and obligations to entities that are not controlled by national governments? If corporation activities did result in environmental damage, how could a dollar assessment be made to the damage of a unique continent – and to whom should the compensation be paid? Whose civil and criminal law should be applied during mining and drilling operations?

Sixth, some form of institution (or institutions) would need to be established to implement a minerals treaty. Should these institutions remain a matter for the Consultative Parties only (bearing in mind that some have no direct involvement in Antarctic mineral matters)? How should decisions be made by the new institutions; should some nations (such as the claimant nations) have some form of veto power? Who would pay for these institutions? Since a mineral regime could result in yet another set of institutions, would it not be more appropriate to develop an integrated treaty for the protection of the entire Antarctic environment which would absorb all the other Antarctic environmental measures?

Just to complicate matters still further, by the time Consultative Parties began work on the minerals treaty, some of the energy anxieties of the 1970s had disappeared. The following appeared in a New Zealand newspaper editorial in January 1983:

> A number of factors have caused countries to reassess their interest in the Antarctic over oil. One is that the demand for oil has decreased. This means that the world's reserves may last longer than previously thought. Another is that a great deal more is now known about the extraction of oil from difficult areas, such as the Antarctic. Much experience has been gained by Canada in the Beaufort Sea and Mackenzie Delta region. One of the lessons learnt from Canada's experience is the enormous cost of mining oil in icy regions. The development costs have been about $300 billion and it has been estimated that about 42 per cent of all Canadian bank lending has been tied up in the projects.
>
> These costs alone make it almost certain that no country or company is going to rush into mining the Antarctic. The fact that the countries which control the Antarctic have not agreed on procedures for mining makes doubly certain that a company or country will not risk the huge amount of capital required for mining while there is some chance that the mining will fail to get the approval of the Antarctic Treaty countries.[7]

Antarctic Minerals Convention

The 1988 Convention on the Regulation of Antarctic Mineral Resource Activities (CRAMRA) is the longest and most complex of the three main

Antarctic treaties examined in this book. But owing to the Australian decision in May 1989, the treaty has no chance of entering into operation, so it is sufficient merely to give an overview of it.

The treaty's stated purpose was to complement the other arrangements that 'ensure that Antarctica shall continue forever to be used exclusively for peaceful purposes and shall not become the scene or object of international discord' (Article 2(1)). 'No Antarctic mineral resource activity shall take place until it is judged, based upon assessment of its possible impacts on the Antarctic environment and on dependent and on associated ecosystems, that the activity in question would not cause. . . .' damage to Antarctica's environment and global weather patterns (Article 4). Decisions about Antarctic mineral resource activities would have had to take into account the need to respect other established uses of Antarctica, including: scientific activities and establishments, conservation of marine living resources, tourism, preservation of historic monuments, navigation and aviation (Article 15).

CRAMRA required the creation of five institutions. The Antarctic Mineral Resources Commission (Articles 18–22) would have consisted of the Antarctic Treaty Consultative Parties, other nations engaged in mineral research and other nations sponsoring such research. This would have been the main political body to oversee mineral resource activities. It would have decided whether or not to open an area for exploration and development.

The Scientific, Technical and Environmental Advisory Committee (Articles 23–27) would have had a membership drawn from all nations party to CRAMRA. It would have advised the Commission; it was not a decision-making body.

Special Meetings of Parties (Article 28) would have been convened specifically to discuss the possibility of opening up an area for mineral resource activity.

A Regulatory Committee (Articles 29–32) would have been established for each geographic area identified by the Commission for mineral resource activities. To service these four institutions a Secretariat (Article 33) would have been established.

CRAMRA distinguished between prospecting ('activities, including logistic support, aimed at identifying areas of mineral resource potential for possible exploration and development'), exploration ('activities, including logistic support, aimed at identifying and evaluating specific mineral resource occurrences or deposits') and development ('activities, including logistic support, which take place following exploration and are aimed at or associated with exploitation of specific mineral resource deposits') (Article 1 (8)–(10)).

The mining system was designed to operate as follows. An 'operator' (for example, Exxon or Shell) had to acquire a 'sponsoring state' (which is a Party to CRAMRA with which it had a 'substantial and genuine link', such as being based in that party's jurisdiction, as with the US or UK). The initial question for the sponsoring state was whether the proposed area of mining was open for that kind of mining. If it was, then an application for exploration would have been filed with the Regulatory Committee responsible for the region in which the proposed mining area fell. If not, then the sponsoring state would have requested the Secretariat to seek approval for opening up the area. A Special Meeting of Parties would have been required. This was only advisory; the key decision would have been made at the next stage by the Commission. Assuming that the Commission agreed, then it would have established a Regulatory Committee: ten members, which would have included the sponsoring state, relevant claimant nations, the USA and USSR. The Regulatory Committee would have taken on the detailed consideration of the proposed mining itself (as distinct from whether or not an area should be opened up for exploration). Assuming that the proposed management scheme was satisfactory, then an exploration permit would have been issued.

Initial reactions to the new convention

The Convention on the Regulation of Antarctic Mineral Resource Activities was adopted on 2 June 1988 in Wellington. It was agreed at the final session that the convention would be opened for signature on 25 November 1988. A special signing ceremony was held in New Zealand for that purpose. In the event, only nine nations signed the treaty at the ceremony: the Soviet Union, New Zealand, Norway, Finland, Brazil, South Korea, South Africa, Sweden and Uruguay. The US, UK, Chile and Argentina said they would sign as soon as possible. Thirty-three nations were eligible to sign it. The delay among these other nations, such as Australia and France, was that each government was divided over whether or not to sign. At least 16 Consultative Parties had to sign and ratify the convention before it came into force – and that number had to include all claimant nations.

Russell Marshall, the then New Zealand Foreign Minister, was pleased with the CRAMRA:

> Regrettably there has been some public misunderstanding about the convention. It does not open up Antarctica for mining. It does not signal the death knell for the Antarctic environment. What it does is put in place agreed and binding legal rules where before there were none. That is the best protection for the treaty system, for the Antarctic environment and for Antarctica as a whole. As far

> as environmental protection is concerned, the convention is an innovative and farsighted document. In fact, in terms of its rigour and its effectiveness, the convention's protection regime has no parallel in international law-making.
>
> I believe that if those who complain about the convention were to read its text more carefully, they would find their fears about potential damage to the Antarctic environment are not well founded. The convention ensures that if ever minerals activities do take place in Antarctica, they will be judged against clear environmental standards, and that agreed means will be in place for monitoring and minimising the impact of such activities and, if necessary, for suspending or cancelling the activities. These are important provisions. They were included in recognition of the fact that Antarctica is a special place and requires special protection. I am confident that this convention does just that.[8]

Mr Marshall's praise was partly justified. CRAMRA did contain some innovations in international environmental law. But if a person were opposed to all forms of mining, then the basic assumption of CRAMRA was flawed: the task of a treaty should be to stop mining – rather than regulate it.

Conservation groups did not, therefore, share Mr Marshall's views. The Washington DC-based Antarctica Project said that the treaty would create:

> . . . a badly flawed regime that will spell disaster if it is ever used to regulate mineral activities. One good aspect of the new Convention, though, is that it provides an internationally agreed moratorium on serious commercial activities, until such time as the member governments agree unanimously to open up a portion of Antarctica to development. Once that door is opened, however, one can have little hope about accountability for the decisions taken, often behind closed doors and without the participation of the public, the press or non-governmental organizations. Our job is to make sure that never happens![9]

Margaret Moore, Australian Conservation Foundation Antarctic Campaign Officer, claimed that CRAMRA:

> . . . has in fact, many severe flaws, most of which will have an adverse impact on the environment. These include inadequate provisions for liability and remedial action for environmental damage, and the confidentiality of data. Under this convention,

> Australia loses all right of veto over mining and cannot prevent mineral activity on any part of its claim.[10]

Conservationist groups began to plan a long campaign to prevent CRAMRA's ratification. Some governments (notably Australia's) were divided within themselves over it. There was, then, some scope for non-governmental lobbying and the chance of eventual success.

In retrospect, I have to admit that I was gloomy about the chances of stopping the treaty; I had been opposed from the outset to its creation. But the bureaucratic momentum built up in true *Yes Minister* style. Too much effort had been invested by too many civil servants for there to be a reversal of the project. Additionally, a form of diplomatic blackmail had developed. It would be 'bad form' for a government, which had been so intimately involved in negotiating the treaty, to decide eventually not to accept it. It is not unknown for governments to reject treaties they had helped draft – the US has done it over the 1982 UN law of the sea treaty. But their participation in the project makes them vulnerable to pressure – as is happening in the US – to accept what they helped create.

A week is a long time in politics. Only six months after the signing ceremony, Australia and then France rejected what they had helped create.

Australia's refusal to sign

The Australian Prime Minister, Bob Hawke, speaking a year after the New Zealand signing ceremony, set out Australia's opposition to the mining treaty:

> The most urgent and relevant action we can take is to ensure that this irreplaceable environment is never put at risk by mining. That is why Australia has decided not to sign the Minerals Convention. This position is based on two simple propositions. First, the Antarctic environment is extremely fragile and critically important to the whole global ecosystem. Second, mining in Antarctica will always be dangerous, and could be catastrophically so.
>
> In the light of those propositions, we are convinced that the Minerals Convention is basically flawed. It is based on the clearly incorrect assumption – current in the 1970s – that mining in the Antarctic could be consistent with the preservation of the continent's fragile environment. But any mining operation, with its accompanying infrastructure and bulk transport needs, would have a lasting and major impact on the area in which it takes place.[11]

Two events helped trigger Australia's decision to block CRAMRA.

On 24 March 1989 the *Exxon Valdez* supertanker, a very modern ship, spilled 240,000 barrels of oil in Prince William Sound, Alaska – the worst oil slick in US history. The oil spill covered an area larger than the size of a small US state, such as Delaware. This is the oil industry's version of Three Mile Island. While other spills like the break-up of the *Amoco Cadiz* off France in 1978 have released far more oil, the Alaskan accident occurred at a time of growing concern among North Americans about environmental damage. More importantly, the concern came as much from middle Americans who could – like the conservative Alaskan fishing communities – see their livelihoods threatened, as from the traditional environmental lobby.

An unqualified third mate was alleged to be piloting the vessel at the time of the disaster. The captain was in his cabin, and blood alcohol tests taken after the grounding indicated he had been drunk. The captain had a record of drink problems and in 1985 underwent six weeks treatment (paid for by Exxon) in an alcohol rehabilitation programme. (For reasons which remain unclear, Captain Hazelwood's record of driving cars while drunk had resulted in his temporarily being without a licence to steer a car – but still licensed to steer a supertanker.) The prosecutors asked for Hazelwood's bail to be set at US$25,000, but State Supreme Court Justice Kenneth Rohl, who had strong views on environmental issues, set the bail at US$1 million (a bizarre decision later overturned on appeal and set again at US$25,000). Hazelwood was accused of piloting the *Exxon Valdez* while intoxicated, reckless endangerment and criminal negligence. The charges, all misdemeanors, were punishable by 27 months in jail and $US11,000 in fines.

In March 1990 Captain Hazelwood was acquitted by an Alaskan jury of the most serious charges but was convicted of a single charge of negligence. The jury rejected charges that he was drunk when the tanker ran aground. Meanwhile, his ship (repaired at a cost of US$30 million) has a new name (*Exxon Mediterranean*), is on a new route (carrying oil from the Middle East to Europe) with a new captain.

The total clean-up cost will exceed US$2.5 billion. Most of the oil has still not been recovered. The oil lost is not hard to locate. It hides in leeward coves and wrinkles in rocks. Thick globules sink below the rocks where they destroy organisms fundamental to the life chain, the green belt. Even where washing took place, the oil survives. A walk across a 'treated' lagoon blackens the boots and kicks up the occasional carcass of a sea otter or bird.

In so far as Exxon would be charged (or paid directly) then the cost would be borne ultimately by Exxon customers who would pay for it through their ordinary purchases of Exxon goods and services. The legal wrangle could go on for years. One hundred and fifty civil lawsuits against Exxon have resulted from the spill. Meanwhile Exxon is suing the

State of Alaska, alleging that the State of Alaska hindered the clean up by opposing the use of oil-dispersing chemicals. Alaska's Commissioner of Environmental Conservation denied this allegation. It appears the industry had failed to stock enough dispersants in south-central Alaska to do a clean-up job effectively. On the night of the disaster the Coast Guard authorized the use of chemical dispersants but Exxon says by then it was too dark and an overnight storm blew much of the oil out of the targeted zone. The case filed by Exxon is a counter-claim to a lawsuit against Exxon and Alyeska Pipe Line Service Company, the industrial consortium operating the trans-Alaskan pipeline and the Valdez terminal. The local environment may be in bad shape – but it is contributing to a healthy legal profession.

Les Hollings, a former editor of the conservative daily newspaper *The Australian* was outraged by what the *Exxon Valdez* meant for the image of big business:

> For Australia, the circumstances of the Alaskan oil spill and the repercussions of it will not only affect our attitude to mineral exploration in the Antarctic but could adversely affect our plans to raise our living standards by our own processing of our vast natural resources.
>
> The extent of the present world move to deregulation will depend on the perceived responsibility of enterprise itself. But where the regulators, the State, have the power to intervene, such as in the Alaskan case, the State lets society down as well.
>
> Development has to continue at the maximum rate possible consistent with responsible environment policies to maintain and improve standards of living.
>
> But if the public feels that there is a nexus between free enterprise governments and irresponsible elements of enterprise then the cause of development will suffer.[12]

Environmental non-governmental organizations were equally outraged – but not necessarily so shocked. After all, they have been warning about the possibility of such a disaster since the early 1970s. The US Congress approved the Alaskan pipeline in 1973 in preference to a longer but safer line through Canada, which would have avoided loading oil on to ships in the sensitive fisheries off southern Alaska. The pipeline owners gave guarantees about protecting the environment, including a commitment to have ample clean-up equipment ready to go into action within five hours of a spill. It was clear that the contingency planning for a spill was based on the assumption that the worst possible accident in the area would release only about one-third of the oil released by the *Exxon Valdez*. In short, the planning was defective.

Additionally, not all the safety precautions were in operation to cope with the tragedy. A spokesperson for the Alyeska Pipe Line Service Company, an eight-company consortium that included Exxon and operated the Valdez oil terminal, conceded that, at the time of the accident, the consortium was in violation of a technical requirement of its own oil-spill clean-up plans, and that those plans may have been unrealistic in any event. The disaster stalled Exxon's chances of obtaining US Government permission to explore for oil reserves in the Arctic National Wildlife Reserve.

The Valdez spill is not the world's worst oil spill. Mexico's Ixtoc well blowout lost 18 times as much crude, and the *Amoco Cadiz* was six-fold larger. Moreover, in other spills, people died; nobody perished at Valdez. The tragedy has become memorable partly because it took place in one of the great unspoiled regions of the globe. It happened in North America, within range of the world's largest concentration of news disseminating organizations. It revealed great human drama such as the battle by fishermen to save their salmon hatchery. The disaster was preventable. The results might have been quite different: if the *Exxon Valdez* had been double-hulled; if the pilot had stayed on board; if Hazelwood had stayed at the bridge; if Alyeska had funded a full-time spill response team; if a storm had not struck when it did.

The disaster was a clear warning of the risk that the 'human element' poses for Antarctic mining. The Prince William Sound waters were far better charted and easier to navigate than Antarctica's. The area was closer to extensive (if inadequate) clean-up personnel and equipment than could be sustained in Antarctica. It is very difficult to clean oil off ice; that would be a primary problem in Antarctica.

In explaining Australia's decision not to sign CRAMRA, Australia's Prime Minister Bob Hawke explicitly referred to the *Exxon Valdez* disaster:

> The Antarctic environment is not only unique, it is especially fragile. A breakdown in some parts of the land, water or air that make up this frozen continent could damage or destroy entire ecosystems.
>
> The grounding of the *Exxon Valdez* in Alaska is testimony to the damage that an oil spill can do to such an environment. In Australia's view, mineral exploitation in Antarctica poses unacceptable risks to the Antarctic environment.
>
> Australia is opposed to mining and oil drilling in Antarctica.[13]

The second development which triggered the Australian Government's decision not to sign the Convention was a state election. Australia took a leading role in drafting CRAMRA but this work attracted little media attention; it also attracted little attention from politicians. In early 1989, however, environmental matters were rapidly proceeding up the national political agenda. A state election in May 1989 resulted in, for the first time ever in any Australian election, five environmental candidates being elected to the Tasmanian Parliament (an increase of three) and holding the balance of power. The world's first 'green' political party was created on this same island almost two decades ago: the United Tasmanian Group. This laid the foundations for the environmental battles over rain forests during the 1980s. Tasmanians were especially sensitive to environmental issues. The conservative state government was replaced by a minority Labor Government kept in power by the five 'green' independent MPs.

Combined with issues such as the popular growing worry about the 'greenhouse effect' and the decline of the ozone layer, politicians in all the political parties throughout Australia suddenly started looking for ways of demonstrating their concern about the environment. Opposing CRAMRA was one such way. Ironically, the then Leader of the Federal Opposition (John Howard) – who had previously expressed little interest in environmental matters, let alone Antarctica – said that Australia ought not to sign. The conservative Federal Opposition, therefore, opposed CRAMRA a few weeks before the Labor Government did so.

France opposes the treaty

The French change of policy was even more surprising. French activities in Antarctica have not represented a notable contribution to its protection. Meanwhile, just north of the Southern Ocean – in the South Pacific – France continues to conduct nuclear testing despite criticism from all South Pacific nations. I remain to be convinced that the new French policy is really a sincere change of mind – and not a comparatively inexpensive way of chasing the French 'green vote'.

Jacques Costeau has been a major campaigner in France. The Costeau Foundation collected a record 1.2 million French signatures on a 'Save the Antarctic' petition.

Costeau's opposition to the mining treaty was partly motivated by the fear of oil spills and other threats to its environment. He also said that:

> The world has no need of these theoretical resources. Known reserves of oil, gas, uranium and heavy metals outside Antarctica are, on a world scale, more than enough to last for many years. And to control the Greenhouse Effect, most experts recommend reducing the use of fossil fuels. Should we needlessly pillage the last unspoilt area of the planet when we have held off exploiting polymetallic nodules, deep thermal springs or shale oil? Given

> these conditions, why plunder Antarctica for raw materials provided today by countries like Zambia, Chile or Zaire, which depend almost totally on their mineral resources for economic survival?[14]

Conservative Australian journalist Padraic McGuinness (who supported the treaty) saw the French decision in a cynical light:

> France has, as has often been remarked, always worked on the principle that it has no permanent friends, only permanent interests. This is, it must be said, a pretty sensible principle to keep in mind in international affairs. So while there is undoubtedly a genuine environmental movement in France (unlike most others, it is honest enough to admit that nuclear power is cleaner than fossil fuel power), and equally undoubtedly a large body of scientific researchers who are committed to research and nothing else, it is perfectly clear that France is maintaining its interest in the Antarctic as much for political and regional defence reasons as for any other.
>
> It is also clear that France's interest in the notion of turning Antarctica into a world park, untouched by human hand, has not prevented it from deciding to continue with its various peripheral activities and to establish a major new base in the interior – a base which will have strategic as well as research functions, like those of the US and the Soviet Union.
>
> Being in the interior on a year-round basis has little other purpose. Not much genuine scientific activity is possible during the worst of the Antarctic night. Except perhaps geological research. What the French researchers do henceforth in the way of research regarding exploitable minerals is subject to no international controls whatsoever.
>
> We do not have an international agreement to prohibit exploration for minerals as a result of the refusal of both France and Australia to sign the Wellington Convention. There is no way in which such activities by France (or anybody else) in the Antarctic continent can be subjected to any regime of international inspection.
>
> There is no probability in the foreseeable future of any international agreement to make Antarctica into a world park, or any other kind of specially protected zone – all we now have to rely on is the Antarctic Treaty, which the Wellington Convention was intended to strengthen. Instead, we now have an unregulated free-for-all as far as mineral exploration is concerned.[15]

McGuinness may be unduly acidic in his comments. France may have made the right decision for the wrong reasons. On the other hand, it is the right decision – and it has helped create the window of opportunity for the environment movement to campaign for an alternative regime.

New Zealand's new policy

Poor old New Zealand. In 1975 it proposed that Antarctica be declared a 'World Park' and got no support.

In the 1980s it chaired the work on creating a minerals regime – and was thus seen (especially after the 1989 decision by Australia and France) as the nation leading the international push for Antarctic mining. As the *Far Eastern Economic Review* pointed out:

> The defeat of the convention by the sudden Australian and French refusal last year to sign the accord they had helped negotiate, and their push instead for the Antarctic to be made a world park, contains a bitter irony for New Zealand. It pursued the minerals convention only when it could find no support in 1975 for a ban on commercial activity on the continent.
>
> Chris Beeby, currently a deputy-secretary of the New Zealand Ministry of External Relations and Trade, who chaired the negotiations, claimed the minerals convention was the toughest conservation regime in the world. It proposed a complex system of defining areas for exploration and mining, with provision for monitoring and strict liability on operators or their sponsoring states to clean up or pay for any damage.
>
> It was, Beeby argued, the best that could be wrung out of countries, including the US and Brazil, which wanted access for their mining companies. New Zealand politicians said when Australia and France pulled out, that failure to implement the convention risked mining on the continent because a world park was unattainable; without the convention there was no protection against commercial exploiters.[16]

New Zealand's policy changed on 26 February 1990. The Prime Minister (Geoffrey Palmer) said that New Zealand had set aside ratification of the proposed minerals treaty and would work for a compromise agreement. Its policy would be based on a series of proposals, including one for a 30-year mandatory moratorium on mining to replace the present voluntary moratorium. Apart from mining, the new approach also embraces protected areas, tourism, waste disposal and marine pollution.[17]

Arani Cuthbert of Greenpeace, Auckland, New Zealand, expressed cautious optimism:

> Greenpeace's response to New Zealand's change in policy is one of cautious optimism. It is certainly a significant breakthrough in the Antarctic campaign but still falls short of a permanent ban on mining and a full environmental convention such as Australia is advocating. If the NZ government really is committed to 'total environmental protection' for Antarctica why don't they embrace a World Park/Wilderness Reserve option?
>
> Palmer's reasoning is that agreement of such a proposal would not be reached by consensus of the Treaty states and 'a creative solution' is needed to break the impasse that has resulted between the World Park proposal and CRAMRA.
>
> Labour's 'creative' solution is a compromise. To defer the ratification of CRAMRA is to put it on the back-burner until the political and industrial climate is more conducive to minerals activity in Antarctica; to announce that it 'may be necessary to look at having a 30 year moratorium on mining' hardly supports the government's claim to oppose mining, it merely forestalls it; to promote a series of separate environmental protective measures for specified problems rather than adopt a comprehensive convention is compatible with re-instating the Convention at a later stage.
>
> Clearly it is a strategy to protect the Minerals Convention by slipping it out of view in the short-term, to protect the present decision making structure – the Antarctic Treaty System ('CRAMRA . . . is now beginning to threaten the Treaty System') and so serve national interests, and to win public favour before an election by appearing to place the interests of the environment paramount.[18]

The UK and US

As the UK's European allies retreated from ratifying the mining treaty, so the UK remained steadfast in its support for it. While France, Italy and Belgium came out explicitly against the treaty, so the UK became the only nation to adopt legislation as though the treaty were going to enter into force. The UK was the only nation to pass a Minerals Act which would allow mining and oil companies to commence commercial operations on the continent. The UK remains interested in the oil potential south of the Falklands – even though it recognizes that such potential would not be commercially viable until at least early next century.

Mrs Thatcher's Antarctic policy was directed at both the emerging public concern over the environment and placating financial opinion. Her public statements contained sentiments in favour of protecting the Antarctic, but she left the door open to eventual mineral exploitation, thereby reassuring companies such as BP, while also knowing that there would be no mineral exploitation during her period in office. This was a wily double-barrelled approach: gaining the support of environmentalists in the short-term, while reassuring the long-term interests of companies.

President Bush has lacked Mrs Thatcher's parliamentary control. The Bush Administration signed the treaty but ratification requires the advice and consent of the Senate. The Administration is unlikely to get that from a 'green' Senate.

At the end of July 1990 Mr Buff Bohlen of the US State Department gave evidence to a Senate Committee enquiry. Like the UK Government, the US Government had to find a formula to avoid alienating both the environmental movement and the financial lobby. The Administration, which was still committed to the minerals treaty, was supporting some form of long-term moratorium. The moratorium could range from 30 to 99 years.

It had resisted efforts by Australia and France to end the agreement in favour of creating a wilderness park where mining would be banned outright. Although the Administration still opposed Australia's plan for a permanent mining ban, Mr Bohlen indicated a moratorium could be designed to have the same effect. 'The difference between a long-term ban and a permanent ban may be slight,' he said, referring to the need for an escape clause to cover the possible discovery of some rare mineral.

> Presumably we are not talking about large-scale mining. We are not talking about petroleum extraction. We are talking about some rare mineral that could be extracted with minimal damage to the environment. Now our proposal would keep that option open. Above all the Administration recognises the need for the parties to adopt much stronger measures to protect Antarctica. We are working with the other parties to supplement the Antarctica Treaty with a new and more comprehensive agreement for fully protecting the Antarctic environment.[19]

The window of opportunity

A cynic about politics and politicians might well suggest that the current concern about Antarctica is only a temporary phenomenon. The general public may become fatalistically resigned to the greenhouse effect and the depletion of the ozone layer. Only time will tell. But this certainly adds force to this book's argument that we currently have a window of opportunity, and that we need to act quickly.

Notes

1. Roger Wilson, 'Nations Meet on Minerals', *New Zealand Times*, 16 January 1983.
2. 'A New Race for the Pole?', *Newsweek* (Sydney) 24 January 1986, p. 127.
3. Barbara Mitchell and Jon Tinker, *Antarctica and its Resources*, London, Earthscan, 1980, pp. 25-6.
4. Barbara Mitchell, 'Undermining Antarctica', *Technology Review*, Boston, February 1988, p. 54.
5. See: Ken Gourlay *Poisoners of the Sea*, London, Zed Books, 1988.
6. 'Antarctic Mineral Resources, Recommendation XI-I', in Department of Foreign Affairs *Handbook of Measures in Furtherance of the Principles and Objectives of the Antarctic Treaty*, Canberra, 1983, p. 1506.
7. 'Antarctic Minerals', *The Press*, 18 January 1983.
8. 'Signing of Minerals Convention', *External Relations Review* (Wellington), No. 1, 1989, p. 33.
9. Circular to supporters dated 16 October 1988 (available from the Antarctic Project 218 D Street SE, Washington DC, 20003).
10. 'Antarctic Minerals Plan and Setback', *Conservation News* (Melbourne) July 1988, p. 7.
11. Bob Hawke 'Antarctica's Future' *The Monthly Record* (Canberra) November 1989, p. 627.
12. Les Hollings, 'Free Enterprise Stained by Alaskan Oil Disaster', *The Australian* (Sydney) 22 April 1989.
13. R.J.L. Hawke, *Our Country, Our Future: Statement on the Environment*, Canberra, Australian Government Publishing Service, July 1989, p. 26.
14. Jacques Cousteau 'Protect Last Continent Against Plunder' *The West Australian*, 11 November 1989.
15. Padraic McGuinness 'France Skates on Thin Ice' *The Australian*, 21 February 1990.
16. Colin James 'Out of the Freezer' *Far Eastern Economic Review*, 22 March 1990.
17. 'Antarctica: NZ Shift' *The Australian Financial Review*, 27 February 1990.
18. Arani Cuthbert 'NZ Government Backs Away from the Minerals Convention' *Antarctic Society of Australian Newsletter*, March 1990, p. 17.
19. 'Antarctic Mining Moratorium Likely', *The Australian*, 31 July 1990.

5. Antarctica on the International Agenda

Outside the Antarctic club

It is now time to leave the club of nations which governs Antarctica to look at the nations and other entities outside the club which want to have a say in Antarctica's destiny. This chapter deals with the efforts inside the UN. The next two chapters deal with a broader range of concerned organizations. There is no central global conspiracy to undermine the cosy club controlling the continent. This means that it is difficult to evaluate fully just how much activity is underway. But it is clear that the momentum has accelerated considerably in the last few years.

The Law of the Sea Conference

The UN's involvement has hinged on the proposed application of the concept of the common heritage of humankind to Antarctica. This concept was popularized by Malta in 1967 in its proposal for internationalizing the sea-bed.

> In view of the rapid progress in the development of new techniques by technologically advanced countries, it is feared that the situation will change and that the sea-bed and the ocean floor, underlying the high seas beyond present national jurisdictions, will become progressively and competitively subject to national appropriation and use. This is likely to result in the militarisation of the accessible ocean floor through the establishment of fixed military installations and the exploitation and depletion of resources of immense potential benefit to the world, for the national advantage of technologically advanced countries.[1]

That came from a four-hour presentation by Arvid Pardo, Malta's Ambassador to the UN, in 1967. The speech, which was long even by UN standards, stunned the delegates since it encouraged them to focus attention on threats to which they had previously paid little attention.

When I met Pardo at the 1975 session of the UN Law of the Sea Conference, he admitted that he was somewhat surprised by all that had flowed from the speech. It remains one of the most important speeches ever made at a UN meeting.[2]

Pardo's use of the term 'common heritage' involved the following key factors. The portion of the sea-bed 'beyond the limits of present national jurisdictions' should be excluded from national control (that is, be subject to some of the international regimes) and there should be some form of international agency to regulate the exploitation of the sea-bed, with the profits going to benefit all humankind (and not just those living in technologically-advanced nations). The regulation and exploitation would have to be carried out so as not to disadvantage those nations already exporting the same kind of minerals to be extracted from the sea-bed. The sea-bed would also have to be demilitarized (and so individual nations could not deploy defence personnel or equipment on the sea-bed). The exploitation would have to be conducted so as to minimize the risk to the marine environment and fishing activities. The proposed sea-bed agency would have to devise ways of co-operating with the innovators in marine technology: transnational corporations. In short, the concept was an immensely ambitious one.

Pardo's speech has three significant implications for this book. First, he popularized the concept of the common heritage of humankind. He suggested that the internationalization of the sea-bed via this concept could avert a scramble by individual nations and corporations for the sea-bed. This book proposes a variation of this concept for Antarctica. A basic difference is that Pardo's concept was concerned about the equitable distribution of sea-bed wealth – the proposed concept of 'public heritage' is concerned only with conservation and would ban all forms of commercial exploitation.

Second, his speech led to the convening of the Third UN Law of the Sea Conference. This chapter reveals a hardly-known initiative in 1975 which, it has been alleged, would – if successful – have seen Antarctica being added to the Conference agenda. Third, the 1975 attempt failed and it was agreed unofficially not to mention Antarctica again in the Conference's context until the new treaty was finalized; the Treaty was finalized in 1982. On schedule, Malaysia raised the question of Antarctica at the UN General Assembly in September 1982.

The law of the sea is one of the oldest branches of international law. Soon after the UN's creation in 1945, it was decided that the UN International Law Commission should bring together the various texts and customs relating to the law of the sea, with a view to codifying and developing them. The work culminated in the 1958 UN Law of the Sea Conference, which adopted four treaties that now form the basis of the law of the sea: the convention on the territorial sea and the contiguous zone, the convention on the high seas, the convention on fishing and conservation of the living resources, and the convention on the continental shelf.

The 1958 Conference was unable to reach agreement on all the contentious questions, most notably the breadth of the territorial sea and the limits to fishing. The Second UN Law of the Sea Conference took place in 1960 and it, too, failed to reach any substantive agreement on these questions. As is often the case, the diplomats and international lawyers agreed to disagree, and hoped that later developments would provide an opportunity to reach agreement.

Arvid Pardo's 1967 statement came at an opportune time. Ocean activities, such as fishing and oil-drilling, had increased considerably in the two previous decades, and the sea-bed in particular was being looked to as a way of providing additional supplies of the earth's non-renewable resources, such as manganese, copper, nickel and cobalt. There was a risk, then, that a scramble for the sea-bed could take place among the technologically developed countries, not unlike that among European countries in Africa in the 19th century – and with the same disastrous results. Pardo urged a halt to national development of the sea-bed, and for the sea-bed to be internationalized and to be mined for the benefit of all humankind. Such a proposal was not welcomed by the technologically developed countries. But the UN is increasingly reflecting the views, not so much of the minority – the wealthy nations – but of the majority: Third World nations. These nations were early supporters of Pardo's proposal.

In 1967 the UN created a committee to study the peaceful uses of the sea-bed and the ocean floor beyond the limits of national jurisdiction. The committee was later enlarged and renamed. What became clear as its work progressed was that it was being hampered by some of the areas of disagreement at the first two Law of the Sea Conferences, for example, what constituted the limits of 'national control'. The UN eventually decided to convene its Third Law of the Sea Conference. The Conference finished in 1982 and was probably the world's largest diplomatic gathering in history. It divided up more of the earth's surface to national control than ever before (about 25 million square nautical miles will go under national control). It had the largest agenda of any conference in history. It handled, particularly in regard to the sea-bed, concepts never encountered before. Its issues cut across customary voting lines so that new and sometimes (by UN political standards) unusual coalitions had to be formed, such as the US and Soviet Union working together to ensure that both superpowers were able, under the new law, to use the oceans and airspaces to deploy their forces against each other. There was also the novel procedure that participants would not call for a vote until the effort to achieve a consensus had been exhausted.

The 1975 Sri Lankan initiative

The Third UN Law of the Sea Conference began its full working sessions in 1973. The Conference soon got bogged down in problems of how precisely a new treaty should be created.

In 1975, Sri Lanka tried to broaden the scope of the Conference's ambit specifically to include Antarctica. The full story of the Sri Lankan initiative is still not known. A set of 'confidential' cables issued or received by the Australian Government arrived anonymously at the offices of the Melbourne-based Antarctic Defence Coalition; details of the cable traffic were made public in May 1980 by the Coalition.[3]

In 1975 Sri Lanka evidently decided that the Conference working group – 'the Evensen Group' – discussing the arrangements for the sea-bed should have its mandate expanded to include the Antarctic region south of Latitude 60 degrees South. A Sri Lankan (the late Shirley Amerasinghe), coincidentally, was then chairing the conference. The main initiator was Christopher Pinto (a Sri Lankan official). His views on the matter were well known: the non-Aligned Nations '. . . would not be satisfied if the development of the Antarctic was left exclusively to the Antarctic Treaty powers'.[4] The 'confidential' cable traffic for August 1975 reveals the mixed reactions that Pinto stirred up.

Three themes in the cable traffic are worth noting. One is the speculation on the extent to which the initiative was mainly Pinto's, or had come from the Sri Lankan Cabinet. The former would obviously be easier to neutralize than the latter. Given the way in which the initiative died – and showed no sign since of being revived – it may have been largely Pinto's.

Second, there is speculation as to who had put Pinto up to making the proposal, and why. The Australian representative in Argentina cabled back a suggestion based on a consultation with a M. Blanco, in Antarctic Affairs at the Argentinian Ministry of Foreign Affairs:

> Blanco also told us (stressing its confidentiality and asking that we did not reveal the source) that Argentina had asked from a 'sensitive source' that Brazil might have been encouraging Pinto through one of their people in Geneva. He asked whether we had heard anything of this: which we had not. It was possible that in view of the difficulties they would have in securing a satisfactory unilateral stake in the continent, Brazil could well have considered its best opportunity for involvement lay in internationalisation.

Such an attitude could encourage Uruguay and Peru to support the initiative also. (Blanco confirmed that this could make it very difficult for Argentina and Chile to carry the Latin group in event of proposed inscription.)

We are unable to assess whether there is anything in the idea of Brazilian involvement, or whether it is just a further manifestation of Argentina's obsessive concern about Brazil. Either way it could present a further complication, giving added important weight to Pinto's initiative, or at least complicating the Argentine approach to a matter where consensus will be important. This might be of interest to you as a background. It would seem from there that even to try to confirm Brazilian interest could risk further development of the initiative.

Finally, there was the secret, international campaign to stifle the proposal. Given the UK's close ties to Sri Lanka, this meant that the UK took on the task of approaching Sri Lanka. Presumably other Antarctic Treaty Consultative Parties would have worked with the other nations with which they had close ties.

The Sri Lankan initiative faded away. An informal understanding was reached that the Antarctic would not reappear on the UN General Assembly agenda until the new treaty was completed.

The Sri Lankan initiative was apparently the only attempt to use the Law of the Sea negotiations as a lever for opening up the political control of the Antarctic. Its failure was indicative that if more nations – let alone all the UN member-nations – wanted to have access to controlling the Antarctic, then an alternative route would need to be found.

1981: India's Operation Gangotri

If a frontal assault would not work, then perhaps more success could be obtained from working within the Antarctic Treaty system. It was India appropriately enough, given its early interest in Antarctica, which led the way.

India was one of the first nations – aligned or non-aligned – to emphasize the need to respect Antarctica's fragile ecosystem. In a memorandum prepared in 1956, India raised this matter at the UN General Assembly; it did not press its views at the time:

Modern science is likely to reveal many possibilities for the peaceful utilization of a region hitherto regarded as unproductive. At the same time the influence of Antarctica on climatic and related conditions throughout the world, while obviously considerable, requires further study. Any disturbance of the

> equilibrium of natural forces in this area might lead to incalculable consequences for the world as a whole involving the deterioration of the conditions for human and other forms of animal and plant life. In view of these facts and bearing in mind the size of the area, its international importance and the growing interest in it, the Government of India considers that in order to strengthen universal peace it would be appropriate and timely for all nations to agree and to affirm that the area will be utilized entirely for peaceful purposes and for the general welfare. All nations should agree further to harmonize their actions to these ends to ensure also that no activities in Antarctica will adversely affect climatic and other natural conditions.[5]

India did not have a research team in Antarctica during International Geophysical Year and so, two years later, was not invited to the international conference which created the 1959 Antarctic Treaty. Ironically, the essence of the concept that India proposed formed the basis of the 1959 Treaty – but India, as far as I can see, received no public credit for its pioneering proposal. Its next Antarctic project – 20 years later – created far more impact.

India's 'Operation Gangotri' was classified 'top secret' and accomplished in record time. 'Not since the dramatic nuclear explosion at Pokharan in 1974 had any [Indian] government operation been conducted with such urgency and silence'.[6] In 1977 Prime Minister Indira Gandhi initiated the project; then other events diverted her attention. Her interest was revived in May 1981 and on 24 July 1981 Operation Gangotri was commenced. Wanting to avoid any risk of coercion by the Antarctic Treaty Parties not to carry out the project, India kept its mission top secret. A Norwegian ship, *Polar Circle*, was chartered and a 21-member team of Indian scientists landed in Antarctica on 9 January 1982. Their principal objective was an investigation of the continent's climatic conditions, especially their influence on the Indian monsoons. A weather station was set up at the camp site to gather further data, which were collected and analysed later. The Consultative Parties were told officially about the expedition a few weeks before the team arrived.

India's Prime Minister said on 24 January 1982: 'We do not subscribe to the view that only a few rich countries have the right to such uninhabited, secluded places. I think we and other countries have an equal right.'[7] India then established the first permanently staffed research station created by a nation that was not at the time bound by the 1959 Treaty.

The dilemma facing India was whether or not to become a party to the Treaty. India decided to become a Consultative Party member, and was admitted to the inner circle on 12 September 1983 (along with Brazil). In other words, India accepted the risks associated with joining the inner club. This meant that it acquired a privileged status but it also created an implicit obligation to help the other Consultative Parties resist the growing Third World pressure to have Antarctica as a UN agenda item. It is interesting that both India and Brazil (the subject of the cable traffic extract above) were admitted within months of Malaysia raising Antarctica at the UN. India and Brazil are Third World nations but each has a gross national product in excess of some of the other Consultative Parties, and both are leading member nations of the Third World. Both also have Antarctic research projects based more on future promise than on past performance.

1982: Malaysia's UN speech

While India was working within the Consultative Party system, Malaysia decided to use the other route: the UN General Assembly. On 20 September 1982, Dr Mahathir Bin Mohamad, the Malaysian Prime Minister, in a wide-ranging speech in the UN General Assembly's plenary session, used the successful completion of the UN Law of the Sea Conference as a peg on which to hang the suggestion that '. . . all the unclaimed wealth of this earth must be regarded as the common heritage of all nations of this planet'.[8] Three months later, at the formal signing ceremony for the Law of the Sea Treaty, Malaysia again referred to Antarctica. On 11 May 1983 the non-aligned conference communiqué called for the 1983 UN General Assembly to study Antarctica.

By the time of the 1983 UN General Assembly session, therefore, nations had been well alerted to the impending clash. Malaysia was undaunted:

> Over the past year the governments of Malaysia and of the small West Indian nation of Antigua and Barbuda have led a campaign to get the issue taken up by the United Nations.
>
> They have come under extraordinary concerted pressure from the United States, the Soviet Union, Britain, Australia and West Germany.
>
> Britain was given the job of persuading the tiny island nation to recant. When Vere Bird, the Antigua and Barbuda Premier, made a private visit to London in July, it was unusual that both Prime Minister Thatcher and Foreign Secretary Geoffrey Howe demanded to see him.
>
> The pressure was counterproductive. It led Third World countries to believe that the treaty powers had something to hide. Antigua government officials say they now suspect that Antarctica's off-shore oil may be close to exploitation, despite

> repeated assurances from the treaty powers that it will not be commercially viable until the next century, if ever.[9]

The eventual resolution,[10] was adopted without a vote – thereby indicating that a consensus had been reached. It was a compromise, with a little for everyone. It kept the subject on the UN agenda but its implementary paragraphs called only for a written report. The Resolution affirmed humankind's interest in the peaceful uses of Antarctica – but there was no explicit endorsement of the concept of the 'common heritage of humankind' in Antarctica's context. The Resolution was less than Malaysia had intended, but at least the issue was now on the UN General Assembly Agenda.

Why Malaysia?

Malaysia's bold involvement in Antarctic affairs was unexpected. Those of us who follow Antarctic affairs expected some nation to raise Antarctica at the UN once the law of the sea treaty was completed. But few, if any, expected Malaysia to break the ice.

A journalist based in Kuala Lumpur noted that 'even the Prime Minister's most avid supporters would accept that his quest for a new Antarctic regime has not exactly gripped the country's imagination'.[11] He would have known that his initiative would be opposed by the Antarctic Treaty Consultative Parties (ATCP) yet not necessarily attract much support from Malaysia's regional allies, the Association of South East Asian Nations (ASEAN). The journalist went on to comment:

> In ensuing exchanges between Kuala Lumpur and ATCP capitals, various political certitudes have been turned upside down: ASEAN solidarity has not been translated into whole-hearted support for the initiative and the two main superpowers – both treaty signatories – have subordinated their rivalry to an unnerving degree, tacitly collaborating in moves to discourage an initiative which both see as something of a nuisance.

Various explanations have been given for Malaysia's unexpected interest. All of them may have some validity.

First, like many non-governmental organizations, Malaysia may have been suspicious of what the Consultative Parties were doing about the minerals regime. If the exploitation of minerals is so far off, why be so hasty? A *Christian Science Monitor* journalist quoted an unidentified Asian diplomat: 'Many of us remember that the Russians sold Alaska to the US in 1867 for $7.2 million. The Czar did not expect Alaska to become a major source of energy for America. We are not about to sell our right to share Antarctica's wealth for a plate of lentils.'[12]

Second, with the gradual erosion of Europe's overseas empires, Antarctica may be viewed by some nations as the most blatant form of colonialism left. The *Christian Science Monitor* journalist went on:

> The question of who owns Antarctica is far from settled by the treaty. Its legal status remains murky. The treaty doesn't acknowledge or reject territorial claims – including some overlapping. The United States and the Soviet Union reject all existing claims, but reserve the right to make claims. The US wants an 'open door' policy allowing free access to the continent's wealth. 'First come, first served,' says one US diplomat privately.
>
> 'The 1959 Antarctica Treaty has no legal value. It reminds one of the 1884 Berlin Treaty through which seven European powers carved up colonial empires in Africa for themselves,' says one African ambassador.

This view certainly underpinned the Malaysian Prime Minister's September 1982 UN speech. In that speech, he said that if there was to be exploitation of resources, the resources would have to be shared beyond the group of developed nations which now administer Antarctica. 'The days when the rich nations of the world can take for themselves whatever territory and resources that they have access to are over,' he emphasized. 'Henceforth, all the unclaimed wealth of this earth must be regarded as the common heritage of all the nations of this planet.'

Finally, the initiative could have been calculated to increase Malaysia's standing among Third World nations. Arvid Pardo of Malta showed, in 1967, how a comparatively small nation can have an impact greatly out of proportion to its size if it selects a key issue at the right time. The 1967 speech met with some opposition from the developed countries, but Malta kept pressing it. In due course, developed nations (except for the US's last minute refusal to sign) gradually recognized that they too could benefit individually from a new treaty. The law of the sea negotiations showed that when a government (such as the US) thinks that its national interest should come before that of the international community, then it will disrupt international negotiations and procedures to follow those interests.

There is no guarantee, therefore, that the unity between the Antarctic Treaty powers can continue indefinitely. Malaysia's 1982 speech may yet prove to be a turning point in Antarctica's political and legal evolution.

UN General Assembly resolutions

The UN General Assembly is the main deliberative body of the UN. All UN member-nations (159 at present) are represented at the annual sessions. General Assembly resolutions are not binding on any nation (except for domestic UN affairs). Their value is derived from their being the best way of assessing governmental opinion on any issue. The General Assembly meets regularly each year from September to late December or January.

On 15 December 1983 the General Assembly adopted – without a vote – a resolution (38/77) requesting the UN Secretary General to 'prepare a comprehensive, factual and objective study of all aspects of Antarctica, taking fully into account the Antarctic Treaty system and other relevant factors' and he was to 'seek the views of all [UN] Member-states in the preparation of the study'. Nations and international organizations involved in Antarctic research were requested to provide the UN Secretary General with 'whatever assistance he may request for the purpose of carrying out the study'. The report was to be ready by the 1984 session, when it would be discussed.

This may seem, at first sight, a very bland resolution. But in the arcane world of diplomacy, Malaysia had achieved a breakthrough. It was less than Malaysia may have wanted. For example, the resolution contained no explicit reference to the 'common heritage of mankind', and calling for a UN Secretary-General study is less than creating a UN Committee to examine – under the UN's aegis – Antarctica's future. On the other hand, the 'Question of Antarctica' was now on the General Assembly agenda. The Antarctic Consultative Parties evidently believed that they could not hope to evade this issue as, eight years earlier, they had at the UN Law of the Sea Conference.

Another way of assessing the impact of the Malaysian resolution is by examining Chile's reaction. Chile claims about 1.3 million square kilometres of Antarctica. The maintenance of that claim owes more to national pride than to a determined scientific endeavour or an international reputation for Antarctic research. Chile is, therefore, a strong supporter of the Antarctic Treaty. Coincidentally, within days of the General Assembly vote, the Chilean Government announced plans to colonize Antarctica with about six 'ordinary families', including young children. 'We don't know of any other nation that is doing this,' said the co-ordinator of the programme, Colonel Hernan Huidobro. 'The purpose is to concretely demonstrate Chile's presence in Antarctica and show that our interests there are physical – not just words on some documents.' Six volunteers – two air force pilots and four civilians – were to take their wives and children to Chile's Lieutenant Marsh base in February 1984. Unlike Chilean military personnel who have been shifting in and out of Antarctica after short solo stays, the pioneer families would stay two years. The plan was to increase the long-term colonists to 20 families in five years. This was done in the context of Chilean news media stories extolling the virtues of Antarctica.

Chile's 'interest' in the region continues. A 1983 special on the government-owned national TV of Chile network showed typical Chilean *huasos*, or cowboys, playing their guitars in front of icebergs, and a folk group singing the music of the Andean highlands while standing in front of a sign pointing to the South Pole.[13]

Each General Assembly session begins with a speech from each delegation setting out its nation's positions on what it regards as the major items on the agenda. This is a useful way of assessing how delegations will be speaking and voting during the next three months or so. The opening speech is often by a head of government or foreign minister, who flies to New York for a few days largely for that purpose.

The Australian delegation to the UN, meanwhile, had become the unofficial spokesperson for the Antarctic Treaty Consultative Parties. It enjoyed a high standing at the UN (where it was a major contributor to UN programmes) and, although allied to the US, it had improved considerably its links with the Third World. On 4 October 1983 the then Foreign Minister, Mr Bill Hayden MP, devoted about 20 per cent of Australia's opening address to advising the General Assembly not to create problems over Antarctica:

> The Antarctic Treaty is a uniquely successful and co-operative international instrument.
>
> It is a major disarmament agreement and excludes all military activities.
>
> It forbids nuclear explosions in Antarctica and prohibits the dumping of nuclear waste.
>
> There is a comprehensive system of on-site inspection, with observers being guaranteed freedom of access at any time to all areas of Antarctica.
>
> The resulting demilitarisation and denuclearisation of the continent – to which the two super powers are parties – is of great value to mankind and of specific value to Australia as a neighbouring continent.
>
> The Antarctic regime is also an instrument of conservation and management of resources.
>
> I should add that the Treaty is not exclusive.
>
> Any state may join and 28 nations with diverse economic and political interests have already done so.
>
> The most recent adherents to the Treaty are India and China.
>
> Australia is concerned about the introduction of this matter into

> the United Nations lest the substantial achievements of the Treaty System be put at risk.
>
> The United Nations was created to solve problems; not to create new ones.
>
> It is the opinion of the Australian Government that any attempt to negotiate a new international agreement on Antarctica or to renegotiate parts of the Treaty would prove counter-productive and introduce uncertainty and possible instability into a region of hitherto unparalleled international harmony.[14]

In the polite, indirect language of diplomacy, that constituted a very blunt statement.

A US magazine following the debate discovered that Australia's Antarctic allies were being equally tough:

> The United States and the Soviet Union, for example, tried to limit dissemination of information on the issue by demanding, successfully, that the United Nations not print a factual account of the Assembly debate on Antarctica in its monthly publication *UN Chronicle*. In addition, the Treaty parties said that they would not participate further in UN consideration of the subject if Malaysia forced a vote to set up a committee that would consider the diplomatic options for the continent. Malaysia's hope was that the Secretariat report on Antarctica would percolate through Third World capitals and generate broader and more active support for its viewpoint, come next year.[15]

On 17 December 1984 the General Assembly adopted without a formal vote a resolution (39/152) expressing its appreciation to the Secretary-General for producing a study and deciding to include this subject on the 1985 agenda. Malaysia had just about managed to keep the issue on the agenda. This was a long way from the proposal, made by the Prime Minister of Malaysia on 9 October 1984, for the UN to establish a special committee on Antarctica to examine the subject in greater depth so as to arrive at a consensus on the objectives of a regime in Antarctica and the machinery to achieve such an objective. The UN resolution again contained no reference to the 'common heritage of mankind'. This was a holding resolution. The Secretary-General's study (to which many nations had contributed) was made available just before the 1984 session; being a substantial document there was not enough time for consideration of it in national capitals.

The year 1985 was a bumper year. Previous Antarctic resolutions were adopted by consensus, that is, without a vote. That year saw the beginning of formal votes, with the Third World using its voting muscle to force through non-binding resolutions.

UN delegates from the Antarctic Treaty nations threatened to boycott debates on the future of the continent unless a group of developing countries abandoned what the delegates described as a confrontational approach. 'Our position at the moment is not to participate in further consideration of the Antarctica item at the United Nations unless and until consensus is restored,' said Australia's Ambassador to the UN, Mr Richard Woolcott, a spokesperson for the treaty parties. He called the UN's lack of agreement 'a matter of great regret to the members of the Antarctic treaty'. Uruguay's delegate, Mr Julio Cesar Lupinacci, said, 'We will not participate in any United Nations decision on Antarctica that is not adopted by consensus.'[16]

The 1985 session adopted three resolutions. Resolution 40/156A requested the UN Secretary-General to produce an updated and expanded study on the 'Questions of Antarctica'. This was adopted with 96 in favour, none against, 11 formal abstentions, with 41 nations having 'announced they were not participating in the vote'. This boycott consisted of about 25 per cent of the UN's total membership. The boycotting nations consisted of the Antarctic Treaty Consultative Parties (minus the missing South Africa) and their immediate allies (such as Afghanistan – then controlled by the USSR – and Israel). Such a large boycott was rare in UN history, perhaps even unprecedented. It was probably the first time that Australia had used this tactic for showing displeasure.

A similar voting pattern occurred over resolution 40/156B, which invited the Antarctic Treaty Consultative Parties to inform the Secretary-General of their negotiations to establish a regime regarding Antarctic minerals. The resolution still did not contain any explicit reference to the 'common heritage of mankind' but it moved closer to it with its affirmation 'that any exploitation of the resources of Antarctica should ensure the maintenance of international peace and security in Antarctica, the protection of its environment, the non-appropriation and conservation of its resources and the international management and equitable sharing of the benefits of such exploitations.' The resolution was adopted with 92 in favour, none against, 14 formal abstentions and 43 nations having 'announced they were not participating in the vote'.

Resolution 40/156C was a bit trickier for some of the formal abstainers and boycotters. The resolution 'viewed with concern' South Africa's continued status as a Consultative Party and urged that the Consultative Parties 'exclude the racist *apartheid* regime of South Africa from participation in the meetings of the Consultative Parties at the earliest possible date'. The vote was 100 in favour, none against, 12 formal abstentions, with 37 nations having 'announced they were not participating in the vote'. India, which had boycotted the previous two resolutions, supported this resolution; China, which had formally abstained in the two previous resolutions, also decided to support this one.

The 1986 General Assembly session also resulted in three resolutions. Resolution 41/88A maintained the momentum of the Malaysian initiative but contained no dramatic breakthrough. It requested the Antarctic Treaty Consultative Parties to keep the 'UN Secretary-General informed on all aspects of the question of Antarctica so that the UN could function as the central repository of all such information'. There was still no reference to the 'common heritage of mankind' and still no decision for the UN to create a separate committee to examine Antarctic affairs. But the resolution's reference to the possibility that the UN '. . . could function as the central repository of all . . .' information from the Consultative Parties represented a potential small expansion of the UN's role.

By comparison with the arcane world of diplomacy, Antarctic glaciers move rapidly.

Resolution 41/88B, on the negotiations for a minerals regime, built upon the previous year's resolution (40/156B). This one called upon the Consultative Parties 'to impose a moratorium on the negotiations to establish a minerals regime until such time as all members of the international community can participate fully in such negotiations'. As noted in the previous chapter, this was ignored by the Consultative Parties.

Resolution 41/88C, on the participation of South Africa in Antarctic affairs, was basically a reaffirmation of the previous year's resolution. This was, of course, a source of embarrassment to some nations (such as Australia) which were trying elsewhere to emphasize their criticisms of South Africa.

All resolutions had to be voted upon. There was no return to the pre-1985 consensus. The voting patterns (including the bizarre boycotts) were similar to those of 1985. Australia (on behalf of the Consultative Parties) and Malaysia had had private meetings before the 1986 session to try to create a basis of consensus. The Consultative Parties, through Australia, indicated a greater willingness to provide information on Antarctic affairs. But Malaysia was obviously seeking more than just information. The moratorium on the minerals regime negotiation (a logical extension of Malaysia's overall views) was not at all acceptable to the Consultative Parties.

The 1987 General Assembly session maintained the pattern.

Resolution 42/46A called for the exclusion of South Africa. Resolution 42/46B once again called for a moratorium on negotiating a minerals regime. Its new idea this time was the request to the Consultative Parties 'to invite the Secretary-General or his representative to all meetings of the Treaty Parties, including their consultative meetings and the minerals regime negotiations'.

Malaysia and its allies were certainly persistent in their efforts. But the Consultative Parties were as adamant as ever. Were the Secretary-General to be invited, then the Consultative Parties would be according recognition to the UN's right to be involved in Antarctic affairs. The issuing of such an invitation, then, would not be merely a polite gesture; in the esoteric world of diplomacy it would have grave implications for enhancing the UN's future role in Antarctic affairs.

The 1988 General Assembly still contained no dramatic breakthroughs. Resolution 43/83A was adopted a few months after the Wellington, New Zealand, negotiations had finished their work on a minerals regime and so the resolution 'expresses its deep regret' that the moratorium requested had been ignored. Resolution 43/83B reaffirmed the request for South Africa's exclusion 'from meetings of the Consultative Parties at the earliest possible date'.

The 1989 General Assembly was the most environmentally active in the UN's history. Fifteen resolutions were adopted dealing with the environment, recommending policies and supporting programmes for action on problems as diverse as floods, drought, large-scale driftnet fishing, pollution by toxic products and hazardous wastes, and the threat of a projected rise in sea-levels due to global warming. Preparations were launched for a 1992 United Nations Conference on Environment and Development. It will aim to bring a global focus to the relationship that binds earth's habitability to economic management. The 1992 Environment Conference, to be held in Brazil, will address nine specific crises of environmental deterioration in the ocean, atmosphere and land resources. It is to produce a range of agreements, recommendations and legal instruments that will get to the root causes of environmental problems and establish responsibilities and obligations for all generations. A centrepiece of the 1992 Environment Conference will be a framework Treaty on the Global Climate, a law of the atmosphere, following on the example of the Law of the Sea though more complicated. A 'framework treaty' would contain basic principles to govern the commonly shared, complex resource that is the atmosphere to which protocols dealing with specific problems (depletion of the ozone layer, acid rain, or any particular pollutant) will be attached as required.

Unfortunately little progress was made on Antarctica. This was foreshadowed on 18 November 1989 by Mr Richard Woolcott, Secretary of the Australian Department of Foreign Affairs and Trade (and previously Australia's Ambassador to the UN). Mr Woolcott noted that his successor at the UN (Dr Peter Wilenski) had called the post-1985 UN General Assembly debates on Antarctica a 'sterile annual ritual'. Mr Woolcott continued:

> This ritual is about to take place again and, regrettably, it seems that there is little prospect of a return to consensus on the basis of the first draft resolution Malaysia has put forward, because, as it now stands, it goes further than previous resolutions in calling for the full participation of all members of the international community in Antarctica. Although it may be a 'sterile ritual', the challenge in the United Nations remains, and will continue to need to be met.[17]

Mr Woolcott challenged Malaysia's claim that the Antarctic Consultative Parties constituted an 'Exclusive Club':

> But there are now 39 governments which have become parties to the Antarctic Treaty. This is a quarter of the membership of the United Nations. It includes a wide range of countries: all five permanent members of the Security Council, developed and developing countries, aligned and not aligned, superpowers and small states. The countries which are parties to the treaty represent, in fact, the majority of the population of the world.
>
> Criticism of the system on the ground of 'exclusivity', therefore, has a hollow ring when such a wide range of countries find it in their interest to join. With the decision in Paris last month to acknowledge Peru, the Republic of Korea and Finland as having consultative party status, the number of consultative parties has more than doubled, from the original 12 to 25 out of 39 states which are now parties to the treaty. We expect other applications for consultative party status to be successful before very long. An institution weakening in the face of criticism does not grow; but the treaty is growing. Nevertheless the challenge will remain and the treaty's capacity to adjust will be tested further as the 90s unfold.
>
> This increase in numbers is welcome, strengthening, as it does, the Treaty System. At the same time, however, it poses a new challenge of its own. The informality of the decision-making process under the Antarctic Treaty, the administrative servicing of larger meetings and the circulation of more documents and other information, as well as the rising level of international interest, put new pressure on the existing rather informal arrangements.[18]

On 15 December 1989 the UN General Assembly adopted two resolutions. In Resolution 44/124A, the General Assembly:

1. Views with concern the continuing participation of the apartheid regime of South Africa in the meetings of the Antarctic Treaty Consultative Parties;
2. Appeals once again to the Antarctic Treaty Consultative Parties to take urgent measures to exclude the racist apartheid regime of South Africa from participation in the meetings of the Consultative Parties at the earliest possible date;

The voting conformed to the now standard pattern. One hundred and fourteen were in favour, none against (South Africa was boycotting the General Assembly), seven nations abstained, five nations were absent – and 31 nations 'announced that they were not participating in the vote'.

Resolution 44/124B contained a small breakthrough: the Antarctic should be declared a 'world park'. The key paragraphs read:

4. Expresses the conviction that, in view of the significant impact that Antarctica exerts on the global environment and ecosystems, any regime to be established for the protection and conservation of the Antarctic environment and its dependent and associated ecosystems, in order to be for the benefit of mankind as a whole and in order to gain the universal acceptability necessary to ensure full compliance and enforcement, must be negotiated with the full participation of all members of the international community;
5. Urges all members of the international community to support all efforts to ban prospecting and mining in and around Antarctica and to ensure that all activities are exclusively used for the purpose of peaceful scientific investigation and that all such activities ensure the maintenance of international peace and security in Antarctica and the protection of its environment for the benefit of all mankind;
6. Expresses its conviction that the establishment, through negotiations with the full participation of all members of the international community, of Antarctica as a nature reserve or a world park would ensure the protection and conservation of its environment and its dependent and associated ecosystems for the benefit of all mankind;

The voting went: 101 in favour, none against, eight abstaining, five absent and 44 nations 'announced that they were not participating in the vote'.

The resolution is notable because of the endorsement of a 'world park' – thereby helping to reduce fears that the Third World's interest in Antarctica is for exploitative purposes. Second, a quarter of the UN membership maintained their, in effect, non-recognition of the UN's role in this matter. Third, by calling for international participation in deciding Antarctica's future, the resolution ran contrary to Australian and French proposals for protecting the Antarctic. The latter were based on the continuing primacy of the Antarctic Treaty system.

The UN's glacier

As at mid-1990, then, the UN General Assembly remains focused upon two clusters of issues: the potential internationalization of Antarctic affairs and the participation of South Africa in Antarctic affairs. The first cluster of issues constitutes the UN's glacier.

The UN General Assembly is, among other things, the home of dormant causes. Sometimes the causes are really lost causes on which no nation can bring itself to impose a 'mercy killing'. This would be done by proposing that the item be struck off the agenda. Instead, an issue will be carried over from one annual session to the next session without any substantive action being taken.

For example, Indonesia invaded the Portuguese colony of East Timor in late 1975 and its troops killed about ten per cent of the population.[19] The UN Security Council called for the withdrawal of Indonesian forces and for there to be an act of self-determination. Indonesia has ignored that instruction. The issue has also been on the General Assembly's agenda, but it is often just deferred from one session to the next; this happened again in 1988 (Resolution 43/402). Indonesia would obviously like the issue to disappear entirely, but it cannot quite muster the votes to ensure this happens. Meanwhile, Indonesia and Australia have decided, quite wrongly in my opinion,[20] to divide up the wealth of East Timor's sea-bed.

By contrast, Namibia was another dormant cause which languished on the General Assembly's agenda. South Africa ignored all requests to withdraw from its illegal occupation of the territory. But the General Assembly continued to reiterate its views. In due course, thanks largely to military and political developments both inside and outside Namibia, South Africa was forced to become amenable. South Africa has now ended its occupation of Namibia.

Will Antarctica languish on the General Assembly agenda as a dormant cause, and gradually slide into the category of a lost cause? Will the Consultative Parties hold out in the hope that an eventual change of government in Malaysia will result in Malaysia dropping its interest in this matter? After all, Sri Lanka did not pursue its 1975 Law of the Sea Conference initiative. Sri Lanka has voted in favour of Malaysia's General Assembly resolutions, but it has never regained its appetite to be a front runner on this subject.

The Consultative Parties have held together very well. The bizarre boycott on voting has brought together a coalition of nations which would generally not be voting together so consistently or so tightly. At first sight, then, they have defused claims that the Antarctic issue is based on the usual configurations of East/West or North/South or Aligned/Non-aligned divisions.

Whether or not the General Assembly will be home for another lost cause will depend upon other factors. These are examined in the next two chapters. But before doing so, it would be useful to look more closely at the role of the spokesperson of the Consultative Parties: Australia.

White Australia

Australia has played a key role as spokesperson for the Consultative Parties. It is also Antarctica's major claimant – claiming about as much of Antarctica as all other nations combined. Australia's claim is about half a century old:

> During [1929–31], British Sovereignty was proclaimed at seven localities, and two years later the Australian Antarctic Territory was established by an Order in Council. This Order was officially gazetted on 16 March 1933, and after the passage of the Australian Antarctic Territory Acceptance Act through Federal Parliament on 13 June that year, it came into force with a Proclamation on 24 August 1936. By provision of the Act, the Australian Government became responsible for the area now known as the Australian Antarctic Territory (AAT).
>
> This territory includes all the islands and territories, other than Terre Adelie, situated south of the 60th degree of latitude, between 45 degrees and 160 degrees east longitude.
>
> Australia's interests also extend to her sub-Antarctic possessions, Macquarie Island, Heard Island and the McDonald Islands.[21]

Indeed, Australia once considered claiming the entire continent on behalf of the British Empire. In 1930 the then Australian Prime Minister, J.H. Scullin, was on the eve of attending an Imperial Conference. R.G. Casey was Australian Liaison Officer attached to the British Cabinet office in London.[22] Casey reminded the Prime Minister of the 1926 Imperial Conference decision:

> As to the Antarctic, you will recollect that at the Imperial Conference in 1926, it was decided that the aim should be the

gradual establishment of British control over the whole (or as much as possible) of the Antarctic continent, apart from the comparatively minor areas to which a good foreign title was held already to exist.

Australia had made a good start:

> Australia has made a good beginning in the shape of the *Discovery* Expedition towards the cementing of our claim to the Enderby Land–Ross Sea area, and the work of *Discovery* in the coming Antarctic summer is likely again to forward our interests.

Unfortunately, Norway was 'our most formidable competitor' and an expedition was underway involving the vessel *Norvegia*. Casey continued:

> In spite of all diplomatic efforts to the contrary, the *Norvegia* 'trespassed' on the western extremity of our area in the vicinity of Enderby Land last Christmas, and actually, I believe, planted Norwegian flags (before the *Discovery*) in the Enderby Land area. Although there has been popular talk of annexation, such action has not been officially proclaimed, and even if it came to this, it is possible that we should be able to counter it, as formal diplomatic warning was given to Norway of our interest in this area before the action on the part of the *Norvegia*. The tone of the Norwegian reply gives ground for thinking that we could sustain an objection to any attempts at annexation on the part of Norway.

Unfortunately for Australia, Norway had a good lead because of its whaling ships:

> However, in view of the pre-eminence of Norway in the whaling industry and their sustained interest in the Antarctic, it would seem to me to be politic to reduce our ambitions (*from being hopeful of getting the whole Antarctic continent*) by voluntarily allotting a sector (say between Enderby Land and Coats Land) in which Norwegian ambition could have free play. [Emphasis added.]

Consequently, Australia had to be satisfied with claiming only about half of the continent. The Australian Antarctic Territory occupies an area of more than six million square kilometres. Today's jet passenger aircraft can fly the 2,000 kilometres from Melbourne, over Tasmania, to the edge of Antarctica in just five hours. It takes far longer to fly from Melbourne to the capitals of South-East Asia.

But Australians themselves have never been very interested in their Antarctic 'empire'. They are unaware of just how large it is (the equivalent of Australia minus Queensland). In so far as they are aware of the golden era of exploration, for example, they are more likely to know about Scott and Shackleton (from the UK) rather than the equally epic work of Australia's own Douglas Mawson.[23] Australia's Antarctic claim is not, then, a matter of national pride. Nor is Antarctic research a major governmental scientific priority – the USSR, for example, conducts more research on Australia's Antarctic domain than does Australia.

If Antarctica were to become a larger domestic political issue, it would be more out of environmental awareness than out of any desire for Australia's flag to fly on foreign soil. There is no strategic reason to be in Antarctica; it would be far more convenient for any potential enemy to attack Australia directly rather than via Antarctica. Many of the early expeditions contained Australians; the scientific work has been maintained. There have been demands for upgrading and expanding Australia's scientific activities, but government expenditure has not met the demands and there is no immediate prospect of a rapid expansion. Additionally, Australia has immense mineral reserves (including oil and gas), and there is no economic argument for an Australian krill industry. In short, Australia has little economic interest in Antarctic exploitation.

Antarctica *is* gradually moving up the Australian political agenda – though its distance from the top should not be overlooked. Its status is changing through four main – and largely separate – debates. First, the scientific community continues to advocate greater expenditure on Antarctic research. Second, Australia's environment movement is co-operating with overseas groups to have Antarctica's environment protected. Third, the peace movement is interested in the Antarctic Treaty as a contribution to arms control and as a model for other nuclear-weapon free zones, such as for the South Pacific and Indian Oceans. Finally, some writers have commented upon Australia's Antarctic claim as an obstacle to developing closer ties with non-aligned nations.

Australia, since the Whitlam Labor Government (1972–75), has moved rapidly to expand its links with the Third World. It is still firmly tied to the United States but has shown of late a far greater willingness to support non-aligned aspirations – and at times to support those aspirations rather than those of the US (especially over the UN Law of the Sea Conference).

Antarctica has two potentials in this context. On the one hand, given increasing non-aligned interest in Antarctica, Australia's claim is an obstacle to improving relations with non-aligned nations. On the other hand, by surrendering that claim in some way, Australia could improve its relations with non-aligned nations more easily than, say, by increasing the scope for those nations to export to Australia (and imperil Australian jobs).

Four publications are worth noting in this context. First, the Sub-Committee on Territorial Boundaries of the Joint Parliamentary Committee on Foreign Affairs and Defence as far back as 1978 warned:

> A challenge to Australia's sovereignty to the Territory by an individual nation is probably unlikely – no country has a better claim if based on early exploration activities. Where the challenge might come from is from the Third World countries questioning the concept of the right of one country to control such an enormous area of an empty continent, particularly as the claim to exercise such control is based neither on occupation nor development but on past historical events. Should Australia lay claim to a 200 mile Economic or Fishing Zone around the coastline of the Antarctic Territory it is possible that such a claim could be contested by countries in the Geographically Disadvantaged group on the grounds that this is making further inroads on resources which should rightfully belong to 'the common heritage of mankind'.[24]

Second, the Fraser Liberal–Country Party Government (1975–83) established a committee of enquiry into Australia's relations with the Third World. In a wide-ranging survey of 14 specific policy areas in which Australia has relations with the Third World, Antarctica received close attention. The report examined Australia's Antarctic involvement and listed three options: strict adherence to maintaining the territorial claim; internationalization; and a middle course based on expanding the 1959 Treaty framework. The committee felt obliged to deal with this issue because it clearly has '. . . Third World dimensions which could become important with time. However, it does not feel called upon to make any recommendations in terms of the options canvassed.'[25] It is notable, none the less, that of all the issues to be examined the committee did at least feel obliged to cover the Antarctic.

Alan Renouf, former Secretary of the Australian Department of Foreign Affairs, writing in his private capacity, was more forthcoming:

> There is another meaningful way in which Australia can improve her relations with developing countries and simultaneously help them materially. Australia claims a large segment of Antarctica. It has become apparent recently that Antarctica is rich in resources becoming scarce elsewhere; before long the technology to tap these resources will be available. There is a growing and reasonable view among Third World countries that these resources do not belong to the claimant countries but should be made available to mankind

> generally through an international regime. Australia has here an opportunity to take an imaginative initiative by becoming the first claimant state to support internationalization of the Antarctic. She has selfishly declined to do so, preferring to sit pat upon her legal claim to an enormous part of Antarctica. This is short-sighted as no other country (not even the US) except Britain recognises the claim or will ever do so and Australia will never be able physically to enforce it. Internationalization of the Antarctic is inevitable but Australia unrealistically refuses to anticipate the inevitable and to benefit from so doing.[26]

Malcolm Booker, a former Australian Ambassador, wrote a study of Australian foreign policy options for the period running up to the 21st century and devoted a few pages to explaining a novel approach to Australia's Antarctic dilemma. He said that little-used provisions of the United Nations Charter for creating trust territories could be drawn upon. The Charter says territories may be governed not only by individual states, but by the United Nations itself. There is also a provision that, where a territory is designated a strategic area, authority over the trusteeship should reside in the Security Council rather than the Trusteeship Council, whose work is now almost ended.

It would, therefore, be possible to make the Antarctic (under Charter's Articles 81, 82 and 83) a strategic trust territory under Security Council control. That would help to maintain the demilitarization of the zone. Another advantage is that four of the states actively interested in the Antarctic (the UK, France, United States and the Soviet Union) are permanent members of the Security Council. It would provide that:

- All states with actual or potential claims to Antarctic territory would permanently renounce them in favour of the United Nations;

- The whole zone, within the 60 degrees parallel, including the seas, would be permanently demilitarized;

- The ownership of all resources in the zone, on land or under the sea, including fish and marine organisms, would be vested in the United Nations;

- An Antarctic commission would be set up, responsible to the Security Council, which would ensure the demilitarization provisions were observed; issue licences to operate in the Antarctic (with authority to deny them to, or withdraw them from, any nation contravening the trust agreement); and co-ordinate and regulate all activities in the zone, including exploration, scientific research and land, sea and air communications.[27]

This is certainly a novel approach – and it would not require any amendment to the UN Charter (which is virtually impossible to achieve). The idea has not yet received attention in the UN General Assembly. Its drawback is that the Security Council has two types of membership: permanent (US, USSR, UK, France and China) and ten nations elected for two-year terms. It is possible that Malaysia and its allies would see this proposal as handing over to the Big Five the main responsibility for running Antarctica.

A new sense of urgency

The significance of the recent initiatives at the UN is that, up until now, Australian Governments have been able to accord Antarctic affairs comparatively little priority. That has now changed. The international spotlight – thanks to non-aligned initiatives – is obliging the Australian Government to give careful consideration to the political and legal future of the Australian Antarctic Territory.

If Australia were ever to be more supportive of UN participation in Antarctic affairs, it would have been since 1983 with the election of a Labor Party Government. The Labor Party has always been more pro-UN than the conservative Liberal Party. It was under a Labor Government (1941–49) that Australia took an active role in creating the UN; by being critical of the US, USSR and UK over the UN Charter in 1945 (which it wanted to be a more ambitious document) Australia was the leading 'Third World' opinion before the Third World was talked of; Australia supplied (in its brilliant Foreign Minister H.V. Evatt) the General Assembly's third President (the only Australian ever to hold that position).

But the Labor Government has done nothing to surrender Australia's claim to its Antarctic Territory nor to boost the UN's involvement in Antarctic affairs. The Government has certainly worked hard to avoid a confrontation at the General Assembly. It was far more tactful as the spokesperson than (say) the US or UK would have been. But there has been a determination to maintain Australia's Antarctic claim.

On the other hand, the Australian Government's May 1989 decision on the minerals treaty has demonstrated its courage in breaking ranks with its Consultative Party colleagues. Further pressure from the environment movement can encourage the Government to go still further.

It is now time to look at the other pressures on all the Consultative Parties.

Notes

1. Quoted in UN Department of Public Information, *The Law of the Sea*, 1973, p. 6.
2. See: Clyde Sanger *Ordering the Oceans: The Making of the Law of the Sea*, London, Zed Books, 1987.
3. 'Conspiracy', *Ice*, Antarctic Defence Coalition, Melbourne, 16 May 1980.
4. Quoted in *The Canberra Times*, 19 September 1977.
5. Quoted in Edvard Hambro, 'Some Notes on the Future of Antarctic Treaty Collaboration', *American Journal of International Law*, April 1974, p. 218.
6. Dilip Bobb, 'Operation Gangotri', *India Today*, 31 March 1982, p. 88.
7. Quoted in, 'An Eye on Antarctica', *Far Eastern Economic Review*, 27 January 1983, p. 32.
8. Text available from Malaysian High Commission, Canberra.
9. 'Third World Wins Antarctic Foothold', *The Bulletin* (Sydney) 11 October 1983, p. 28.
10. UN General Assembly Resolution 38/77, 15 December 1983.
11. 'Breaking the Ice', *Far Eastern Economic Review*, 1 December 1983, p. 25.
12. Louis Wiznitzer, 'Antarctica: High Stakes in Icy Diplomacy Tussle', reprinted in *The Sydney Morning Herald*, 23 May 1985.
13. 'Chile to Colonize Antarctica in Bid for Sovereignty', *The Australian*, 24 December 1983.
14. 'Foreign Minister's UN Speech', *Ministerial Statement – Department of Foreign Affairs*, Canberra, 4 October 1983.
15. 'Challenge to the Antarctic Treaty Goes Nowhere', *The Interdependent* (New York) November 1984.
16. 'Boycott over Antarctica Threatened by Delegates', *The Sydney Morning Herald*, 6 December 1985.
17. Richard Woolcott 'Antarctica's Future: The Challenges and Changes' *The Monthly Record* (Canberra) November 1989, p. 630.
18. Ibid., pp. 630-31.
19. See: Carmel Budiardjo and Liem Soei Liong *The War Against East Timor* London, Zed Books, 1987.
20. Keith D. Suter, 'The Timor Gap: Oil in Troubled Waters', *Inside Indonesia* (Melbourne) December 1988, pp. 7-9.
21. M.S. Betts, *Australians in Antarctica*, Canberra, Australian Government Publishing Service, 1981, p. 16.
22. Neville Meaney, *Australia and the World: A Documentary History from the 1870s to the 1970s*, Melbourne, Longman Cheshire, 1985, pp. 374-5.

23. Lennard Bickell, *This Accursed Land*, Melbourne, Macmillan, 1977.
24. Joint Parliamentary Committee on Foreign Affairs and Defence – Sub-committee on Territorial Boundaries, *Australia, Antarctica and the Law of the Sea: Interim Report and Appendices* Volume I, May 1978, p. 73.
25. *Report of the Committee on Australia's Relations with the Third World*, Canberra, Australian Government Publishing Service, 1979, p. 169.
26. Alan Renouf, *The Frightened Country*, Melbourne, Macmillan, 1979, p. 526.
27. Malcolm Booker, *The Last Quarter: The Next Twenty-five Years in Asia and the Pacific*, Melbourne University Press, 1978, pp. 118-19.

6. Coming in from the Cold

Antarctica's new era

Antarctica is coming in from the cold. The private cosy club of 1959 is now undergoing international scrutiny. Even if the Antarctic Treaty Consultative Parties manage to continue to block the UN General Assembly's involvement in Antarctic affairs, there are other signs of the club's control over Antarctica being eroded.

The next three chapters trace Antarctica's increasing importance in international politics. This chapter begins with South Africa's participation in Antarctic affairs. Antarctica is now one of the few areas in which South Africa is not ostracized, but its participation is attracting unwelcome attention to Antarctica. The French participation is not nearly as controversial but it, too, has attracted unwelcome attention. An emerging issue in Antarctica is the disposal of waste generated by people who, ironically, are doing research there. It could be assumed that they in particular would be taking care of the waste generated by their bases. Recent evidence indicates that this is not always the case.

Chapter 7 deals with Antarctica's presence in the debate over protecting the environment. Antarctica has received attention via independent commissions which have been established to provide comprehensive programmes for international action on global problems. It summarizes two publications, published at either end of the 1980s, which call for a more internationalized management of Antarctica's affairs. Antarctica is now also receiving attention because the gap in the ozone layer was first detected above that continent. A further current worldwide environmental worry is global warming – the greenhouse effect – and Antarctica receives attention in this context because of speculation that global warming may lead to a melting of Antarctica's ice and consequent coastal flooding around the world.

Finally, this book has contained references to the work of non-governmental organizations (NGOs). These have helped, for example, to bring about the 1989 changes in Australian and French policy. Chapter 8 explains the significance of NGOs and then looks at the NGO work on Antarctica.

Taken together, then, Antarctica is entering a new era. For various reasons, Antarctica is becoming a major political issue, mainly – but not exclusively – among people and organizations concerned with the environment. The views being expressed by all these parties are far from uniform. But in total they do represent a determination to open up Antarctica's affairs to greater international scrutiny and involvement.

The South African connection

> The *apartheid* system denigrates and insults the dignity and worth of black people not only in South Africa but all over the world and it is this expression of racism which constitutes a threat to the peace. Over thirty years ago, at Nuremberg, it was solemnly declared that racism constitutes a crime against humanity. If racism then constituted a crime against humanity and a threat to world peace then *apartheid* today is equally a crime against humanity and presents perhaps an even greater threat to world peace.
>
> The very nature of the *apartheid* system with the brutal and violent imposition of white domination upon the majority population has created a highly explosive situation which is likely to erupt into a major racial confrontation with the danger of a wider war which will set aflame the continent of Africa.[1]

South Africa's *apartheid* racial system began its regular appearance on the UN General Assembly agenda in 1962. (South Africa ratified the 1959 Antarctic Treaty in 1960.) Previous attempts to raise *apartheid* had been defeated on the grounds that this was a 'domestic matter' and the UN Charter (Article 2(7)) prohibited UN involvement in matters which are essentially of a domestic nature. By 1962, however, it was clear that South Africa's racial policies were not just a matter for South Africa alone. As the years rolled by, African suffering increased and racism in the world generally became far less socially acceptable, so South Africa's public allies dwindled. It has continued to enjoy trading arrangements with nations which are publicly critical of its racial policies, but no nation now defends South Africa's *apartheid* policies.

The previous chapter showed how the Malaysian initiative on Antarctica soon mutated into various sub-initiatives as delegations learned more about Antarctica. UN General Assembly sessions are always a useful way for governments to learn new information. South Africa's participation in Antarctic affairs was not a secret – but nor was it an event that its Antarctic colleagues would go out of their way to publicize; they realized how embarrassing their association with South Africa could become.

South Africa's Antarctic status contains four paradoxes. First, that country is being ostracized in almost all other international organizations because of its *apartheid* racial policies. However, there has been no attempt to discourage its involvement in the political meetings associated with Antarctica's affairs. Second, South Africa has not been very active in Antarctica's scientific affairs for some years but remains a Consultative Party – while other nations, aspiring to be Consultative Parties, have to prove some serious scientific activity in Antarctica. Third, Antarctic meetings represent an important back-door way for South Africa to maintain its international contacts, thereby evading the international boycott which some have urged against South Africa. Fourth, some nations which are publicly critical of South Africa are obliged, however, at each annual UN General Assembly session to boycott the vote on South Africa's involvement in Antarctic affairs. Richard Woolcott, then Australia's Ambassador to the UN, said:

> The Australian Government has unequivocally condemned and deplored the institutionalized racism, which the system of *apartheid* in South Africa involves. But the Antarctic Treaty is a functioning international instrument, which cannot be replaced or altered except by its contracting parties. In this connection I should say, with reference to South African membership of the Treaty, that this is really an issue for the Parties to the Treaty themselves to determine, as is the case in any other Treaty system.[2]

South Africa's participation in meetings of the Antarctic Treaty Consultative Parties will continue to attract criticism at annual sessions of the General Assembly. Criticism of *apartheid* is one issue upon which most of the UN's membership agrees. For this reason alone, then, Antarctica will remain on the General Assembly's agenda until either South Africa ceases to be a Consultative Party or it ends *apartheid*.

The most recent South African Antarctic activity which generated controversy was based on reports that South Africa is planning an airstrip on Marion Island and that this may have been part of its preparation for conducting a nuclear test there. The project was stopped, thanks largely to domestic environmental pressure. But it was another reminder of South Africa's unfortunate ability to draw adverse publicity to itself – and, by implication, to embarrass its fellow Antarctic Treaty Consultative Parties.

South Africa was one of the first nations to become involved in the nuclear fuel cycle since the UK hoped it would become a source of uranium for British nuclear requirements. During the 1950s, South African universities expanded their research into nuclear physics. South Africa is an important player in the international nuclear fuel cycle, though not all of its activities are easily monitored.[3] It is both selling uranium and developing its own nuclear power programme.

One component of its nuclear power programme is the alleged development of nuclear weapons. As long ago as 1978 there was speculation about its nuclear ambitions:

> At some future date nuclear weapons may be introduced into southern Africa. In fact, in August 1977 the United States and the Soviet Union were convinced that South Africa was about to conduct a nuclear test explosion. Strong diplomatic pressures were applied to prevent this. Whether or not South Africa then intended to explode a nuclear device, there is no doubt that it could do so rapidly once the political decision is taken . . .
>
> Why would the South African regime want nuclear weapons? The use of these weapons in southern Africa would make no sense, although some extremists may argue for it as a last-ditch effort. But the policies of the United States and the Soviet Union in the area, whether they are based on influencing events directly or by proxy, are bound to be affected by the presence of nuclear weapons. The South African regime may, for example, even imagine that the threat to use nuclear weapons may at some future date force the United States to support it out of fear that such use might escalate to a general nuclear war between the United States and the USSR. The Pretoria Government may also believe that its strategy of creating buffer States to the north may in some way be assisted by the acquisition of nuclear weapons. Finally, a regime in desperate straits tends to use all the means at its disposal in an effort to survive.[4]

South Africa's first nuclear test may have taken place on 22 September 1979. A US spy satellite recorded what could have been a small nuclear explosion in Namibia (South West Africa). The evidence for and against the explosion having taken place is still not definite.[5] With the independence of Namibia, this has stopped being a testing area for South African nuclear devices; such testing will need to be done elsewhere – hence the interest in Marion Island.

Marion Island, part of the Prince Edward Islands group, is at 46 degrees South. It is, then, outside the geographical area of application of the 1959 Treaty (which starts at 60 degrees South) and so Consultative Parties cannot insist on inspecting it. But it is inside the Antarctic Convergence and so covered by the Convention for the Conservation of Antarctic Marine Living Resources (to which South Africa is a party). But that Convention contains no provisions of its own regarding nuclear and military activities. If South Africa were to engage in such activities on Marion Island, it would therefore technically not be in breach of its obligations under the Treaty or under CCAMLR.

South Africa has operated a scientific research station on Marion Island since 1948; the station is part of the South African National Antarctic Research Programme. In December 1986 a British newspaper reported that South Africa intended to build an airstrip on Marion Island. This was of concern to South African conservationists because of the potentially adverse impact it would have on the environment.

Ms Nan Rice, of the South African Dolphin Action and Protection Group (DAPG), has written an account of subsequent events.[6] The South African Department of Environment Affairs (DEA) was in the process of commissioning an environmental impact assessment (EIA) around the time of the British press speculation. The DEA said that an airstrip had been proposed, but no mention was made of there being a missile or a nuclear testing facility. The DEA had decided to commission an EIA before work commenced (something, incidentally, the French have not done on their Antarctic territory). It would be an open investigation, and would be subject to an independent review by a British expert (who was subsequently barred by the British Foreign Office from taking part – presumably as a result of official British reluctance to be implicated in this controversy). The EIA advised against the project and the project was then cancelled. The South African *Cape Times*, in an editorial, stated that: '. . . the South African government decision to scrap plans to build an airstrip on Marion Island is, in election parlance, a gain for the conservationists'.[7]

Ms Rice was critical of the press speculation and the role of foreign conservationist non-governmental organizations in fuelling a controversy which was, in fact, resolved satisfactorily by domestic South African procedures. I have sympathy for the DAPG; it must be annoying for South Africans to have their government constantly subject to the worst possible interpretations of its actions. But that is the automatic by-product of being internationally ostracized for having racial policies which no other government now endorses. Meanwhile, South Africa still has a nuclear programme and will remain subject to international scrutiny. The Marion Island controversy is bound to be repeated in some other context. South Africa's Antarctic involvement will continue to be controversial.

Meanwhile the Antarctic Consultative Parties are presumably hoping that President de Klerk will be able to introduce sufficient reforms to enable the UN to reduce its criticism of South Africa. Some steps have been made in the right direction, such as the release of Nelson Mandela. But, given the strength of the white opposition to the type of real reforms that still need to be made, President de Klerk will not be able to dismantle *apartheid* easily. South Africa will, then, remain a constant embarrassment to its Antarctic allies.

Indeed, domestic political difficulties for these allies may get worse. As

South Africa becomes still more polarized and violent, so critics of *apartheid* based in Australia, the UK, New Zealand and the US, may pay increased attention to the inner workings of the Consultative Parties as a way of trying to keep the pressure on South Africa to reform.

The French connection

France is another Antarctic Consultative Party whose actions are attracting adverse international scrutiny. One action is the construction of an airstrip at Pointe Geologie, an archipelago near the French base at Dumont d'Urville. This involves the flattening of five small islands, which are the habitats of several species of Antarctic birds, some unique to that region. Moreover, the construction has involved the deaths of considerable numbers of penguins and other birds. Both actions – the death of the birds and the destruction of habitats – are specifically prohibited under the provisions of the Agreed Measures.

The other action, which is still in the realm of speculation, is the commencement of French nuclear testing on Kerguelen Island.

A growing problem in Antarctica is the competition between humans and other living creatures for ice-free areas. Only about two per cent of the continent is ice-free all year round. That is an area about the size of Britain. The French civil engineering project is an example of this scramble for rock.

The French wish to extend their scientific season and for this they need access for supplies to arrive. Sea access is viable for only about two months each year. Their solution, begun in 1983, is the construction of an airstrip. Explosives were used to move rocks and earth; these killed at least 25 penguins and over 1,000 eggs. The work was halted in 1984–85 in response to international criticism. Alternative ways of conducting this work were investigated. The work has since been resumed but apparently without the explosives. The airstrip will, however, have long-term adverse environmental consequences.

Environmental non-governmental organizations such as the Antarctica Project and Greenpeace International have argued that the airstrip is a breach of the rules laid down in the 1964 Agreed Measures drawn up by the Consultative Parties for the Conservation of Antarctic Fauna and Flora. The Agreed Measures provide a basic structure of protection for all native species by forbidding the killing of mammals or birds, the creation of Specially Protected Areas and by the creation of Sites of Special Scientific Interest.

The Antarctica Project and Greenpeace International have stated:

> The wording of the Agreed Measures, unlike other Recommendations of the Antarctic Treaty System, is explicit. Disregard for

> the regulations is permitted only in the event of an 'extreme emergency involving possible loss of human life or involving the safety of ships or aircraft'.
>
> Article VI of the Agreed Measures prohibits the 'killing, wounding, capturing or molesting of any native mammal or native bird, or any attempt at any such act, except in accordance with a permit'. Permits are given only to provide indispensable food, or specimens for scientific work, museums or other educational or cultural institutions. The French had no such permit. The killing and wounding of penguins during a construction project are not among the exceptions listed in Article VI.
>
> Article VII of the Agreed Measures requires the minimization of 'harmful interference within the Treaty Area with the normal living conditions of any native mammal or bird . . . ' Included in the list of activities considered to be harmful is the 'use of explosives close to concentrations of birds and seals'. Harmful interference is, however, permitted by Article VII, 'to the minimum extent necessary for the establishment, supply and operation of stations'.
>
> The French government has used this latter provision to justify its activities in constructing the airfield. But Dumont d'Urville has been established, operating and supplied for many years. Clearly the airstrip is not 'necessary' for any of these to continue. Even if it were determined that the airstrip is 'necessary' for the supply or operation of the base, the project underway cannot be seen as 'minimizing' the harmful interference.[8]

France's fellow Antarctica Consultative Parties have been unable or unwilling to stop the French activities. The work was commenced before an independent environmental impact assessment was made (South Africa's Marion Island episode was better in this respect). The entire saga is a poor omen for the Consultative Parties to be able to claim to enforce environmental controls among themselves. Indeed, the Consultative Parties seem to have done little (if anything) to exercise their right under the 1959 Treaty to even investigate just what the French are doing. In terms of drawing international attention to this matter, the non-governmental organization Greenpeace has itself done the work of the Consultative Parties. Greenpeace has investigated the project and carried out a demonstration on the site. Greenpeace has now produced the photographic evidence of the destruction which has been used by the world's media. All this reflects badly on the Consultative Parties.

But France may have even more shocking things in store for its Antarctic Treaty partners: nuclear testing.[9] The first French atomic device was exploded in 1960 in the Sahara. During the next two years, France had four atmospheric tests and thirteen underground tests.

Algeria became an independent State in 1962 and France decided to follow the US and British examples of using the South Pacific for its tests. This was presumably done in the belief that the tests would attract little adverse international attention, for the South Pacific was a considerable distance from the world's politically most important countries, and the population in these mainly less developed countries (some of them still colonies), could put up little effective opposition to the tests. Mururoa, a small atoll in French Polynesia, now holds the region's record as host for the largest number of nuclear tests, and is probably the most heavily polluted island in the Pacific.

One reason, then, for assuming that the French would be interested in an alternative test site is that the current site is now too damaged to sustain many more tests. Mark Delmage has quoted Abraham Behar, a French biophysicist at the University of Paris: 'France will abandon Mururoa because it is so leaky and contaminated . . . The need to place recent tests deeper into the island's deteriorating volcanic base had made it technically impossible for many more blasts to be conducted.'[10]

The second piece of circumstantial evidence comes from the gradual move towards independence of French colonies in the Pacific. It was, of course, the independence of the French colonies in North Africa which prompted the transfer of testing to the South Pacific. In late 1984 the French Consul-General to Victoria, South Australia and Tasmania, Bernard Ledun, said that should French Polynesian territories soon gain independence, it was likely France would have to relocate its nuclear testing site from Mururoa, 7,000 kilometres east of Australia in the Pacific. 'My personal view is that the ultimate relocation for our testing could be the Kerguelen Islands,' Mr Ledun said.[11]

A third piece of circumstantial evidence is the growing regional opposition to French nuclear testing. Three regional nations (Australia, New Zealand and Fiji) took France to the International Court of Justice in 1975 in the hope of forcing an end to the tests. The 1985 French destruction of Greenpeace's *Rainbow Warrior* contributed to the region's perception of French ruthlessness. Also in 1985 came the South Pacific Nuclear Free Zone Treaty drafted by the thirteen independent nations of the South Pacific. One provision was aimed at ending French nuclear testing. France has refused to sign this Treaty – a decision copied by its two nuclear allies, the US and UK. Opposition to nuclear testing is one of the few policies upon which all the South Pacific nations agree.

The final piece of circumstantial evidence comes from French sensitivity about foreign inspections of Kerguelen Islands, which are part of the French Southern Ocean and Antarctic Territories. One of the earliest examples of speculation about French nuclear plans for the Kerguelens arose in 1980 when the French Minister of Defence, who had just been inspecting the Mururoa test site, refused to confirm or deny that there were plans to move the test site to the Kerguelens. Since France was running out of overseas territories, its range of options was narrow and presumably the Minister did not want to rule out any potential site. The speculation has persisted.

On 9 October 1986 the French navy sank a fishing trawler, *Southern Raider*, near Kerguelen Islands. Skipper John Chadderton and first officer Alistair Annandale had no doubt that its sinking had some connection with Kerguelen. In an interview, Chadderton said, 'They accused us of illegally fishing in French waters, but they interrogated us endlessly about espionage, not fishing. During the three months we were held by the French on Reunion Island we became convinced that France was planning to move its nuclear tests to Kerguelen. This is commonly accepted as fact on Reunion, the French administrative centre closest to Kerguelen.'[12]

Senator Jo Vallentine, Australia's first Independent Senator for Disarmament, questioned the Australian Government:

> What evidence has the Minister representing the Minister for Foreign Affairs got that the sinking of the *Southern Raider* on 9 October and the detention of four Australian and four New Zealand crew members is nothing to do with illegal fishing for lobsters or drug smuggling but is due to French fears of spying? Is it true that the French Government is preparing a nuclear test site on Kerguelen Island, which is about 3000 kilometres south-west of Perth and from where the prevailing winds blow directly to Australia? Has the Australian Government made any representation to the French Government objecting to the construction of a nuclear test site at Kerguelen, which is close to the Australian islands of Heard and Macdonald and much closer to Australia than Mururoa Atoll? Is the Government prepared to challenge the French authorities to tell the truth about the sinking of the *Southern Raider* and why the crew has been detained on Reunion Island so that we do not get a repetition of the lies and falsehoods which followed the sinking of the *Rainbow Warrior* in July 1985?[13]

The Government's response, via Senator Gareth Evans, was ambivalent:

> The captain and second-in-command of the *Southern Raider* were charged on 17 October with infringing French fishing and merchant navy laws. Subsequently they were placed under judicial control for infringement of drug laws. Magistrates on Reunion are collecting information on the sinking of the *Southern Raider* to determine

> whether to proceed to trial. We therefore have no evidence at this stage that the detention of the crew of the *Southern Raider* is related to French fears of spying. It will not be until the magistrates decide whether to proceed to trial – we have a full list of the formal charges against the crew – that we will be able to determine whether representations to the French Government are appropriate. We have asked the French authorities to finalise the legal processes as quickly as possible so that we can be in a position to respond appropriately.
>
> The question of possible French interest in nuclear testing on the Kerguelen Islands first arose in December 1984 when the French Consul-General in Melbourne, speaking in his personal capacity it seems, mooted this possibility. The French Defence Minister at the time, Charles Hernu, denied any such plans and the Australian Government has no reason to doubt this denial. This remains the case today. The Government is not aware of any evidence that would lend support to more speculation about French interest in nuclear testing on Kerguelen. I should just finally add that Australia's strong opposition to French nuclear testing in particular stems from the fact that France is the only country that does conduct nuclear tests outside the home territory of a nuclear weapons state and we have repeatedly insisted that, at the very least, France should take its nuclear test programme home to metropolitan France.[14]

The Kerguelen Islands are about 14,000 kilometres from Paris and 3,000 kilometres south-west of Perth, in the southern extremity of the Indian Ocean. The archipelago is made up of one main island *Grand Terre*, surrounded by hundreds of small islands with innumerable islets and rocks. It totals 7,215 square kilometres, has an extremely rugged coastline with the highest point being Mount Ross at 1,850 metres. Glacial lakes, glaciers which reach right down to the sea, hot springs, boggy peat marshes are to be found – the rugged terrain has ensured that the interiors are not well explored. The whole coast is broken up by deep fjords bounded on either side by perpendicular cliffs, and the region is swept by great westerly winds – an average of a gale a day.

The islands, geologically speaking, may be only marginally 'better' than Mururoa for nuclear testing. They would not be as popular an accommodation area for service and scientific personnel involved with the tests. But the islands as a testing site would enable France to evade some of the political problems in the South Pacific.

Radioactive leakage from Kerguelen Islands would be carried by wind and sea to Australia. The islands lie in the path of the Roaring Forties and the Furious Fifties, and prevailing westerly winds which flow across the Southern Hemisphere at between 40 and 50 degrees latitude. Ocean contamination could be brought to Australia by western currents which sweep along the west and south coasts of the continent. Radioactive debris would contain strontium 90 which has a half-life of 28 years. This constitutes a major health hazard, causing bone cancer and leukemia. Brenton Groves, writing to the *Canberra Times* in 1982, set out other problems for Australia:

> There is no doubt that France owns Kerguelen Island but Heard and McDonald Islands, which legally come under ACT [Australian Capital Territory] law, are only 361 kilometres away. Australia has declared an EEZ (Exclusive Economic Zone) of 200 nautical miles where 'the State (in this case Australia) will have sovereign rights for the purpose of exploring and exploiting not merely the natural resources of the continental shelf but also the resources of the superjacent waters as well'. Therefore, a direct legal conflict with France exists.
>
> Exploding nuclear weapons underground in Kerguelen Island has two direct effects on Australian property. Both Kerguelen and Heard Islands are integral parts of the Kerguelen-Gaussberg ridge, which extends down through several sea mounts to the Antarctic territories claimed by Australia. We do not know enough about the geological formations in that area of the world, having done very little research, but one thing is sure. Every bomb that goes off in Kerguelen Island will be felt in Heard Island as well as over a wide range north and south of both.
>
> The encyclopaedia says that on these islands seafowl such as penguins, albatrosses, petrels, are numerous and that the surrounding waters abound in fish, whales and seals. This is in contrast to the domestic fishing industry but, again, we do not know how rich the waters around Heard Island are, due to lack of research. However, Antarctic waters are part of an extensive food chain that cannot possibly be helped by heavy doses of radiation.[15]

Australia, therefore, has a close interest in not having French nuclear testing conducted on the Kerguelen Islands.

The most obvious treaty which could have a bearing on this matter is the basic 1959 Antarctic Treaty.

Article V bans the testing of nuclear weapons or the storage of nuclear waste in the Antarctic. Unfortunately the Kerguelen Islands are outside the geographical area of application of the Treaty (south of latitude 60 degrees South). One by-product of this exclusion is that whereas bases on the territory covered by the treaty are open to inspection by all Consultative Parties, no non-French citizen has any right to try to inspect

French bases and activities on the Kerguelen Islands. French military activities on these islands are, in effect, as secret as those conducted on (say) a base in Paris. This helps explain the Australian Government's vagueness over the *Southern Raider* incident: Australia has no automatic right of inspection.

The Kerguelen Islands do, however, lie within the geographical area of application of the 1980 Convention on the Conservation of Antarctic Marine Living Resources (CCAMLR). As described in Chapter 3, this followed an 'ecosystem' approach – rather than the lawyers' preference for straight lines – and uses the Antarctic Convergence as its outer boundary. Unfortunately, CCAMLR itself does not contain any provision relating to nuclear activities. France is not breaking the letter of the CCAMLR Treaty.

Ironically, the Kerguelen Islands received an explicit mention in the CCAMLR Conference. The Kerguelens are different from the Antarctic claims. There is no overall agreement about the claims; each is recognized by only some nations. The Kerguelens, however, are recognized as French territory. France wanted the Kerguelens to receive separate status. A Declaration was included in the Final Act (that is, final document) of the CCAMLR Conference:

> French conservation measures adopted before the entry into force of the Convention would remain effective until modified by France, and conservation measures under the Convention can only be applied to the waters adjacent to the islands with French consent. France could also adopt measures which it might deem appropriate, and would carry out national or international enforcement in those waters. The observation and inspection system may only be applied as agreed by France and in the manner so agreed. No objection was made to the statement.
>
> The legal effect of this statement is small . . . Of far greater concern are the political implications of this legally redundant and ineffective recital. France has unambiguously indicated that it is prepared to prevent Convention conservation measures applying within the EEZ [exclusive economic zone] of its sub-Antarctic islands, thus considerably reducing the ambit of the ecosystem approach.[16]

Although France can claim that nuclear testing would not break the specific wording of the Antarctic Treaty and CCAMLR, it would, in my view, contravene the spirit of the treaties. This is not a sensible period to be breaking the spirit of treaties.

Breaking the spirit of both treaties would demonstrate what the Third World nations are claiming – that the old guard cannot be trusted to act responsibly in Antarctica. It would create a tougher coalition between environmental and peace non-governmental organizations and Third World nations in seeking to break up the cosy relationship. It would add to the internal tensions within the Consultative Parties – some of which are already embarrassed by having to permit the continued involvement among their number of South Africa. Australia, for example, defends South Africa's continued status as a Consultative Party; would it need to add to that list of headaches the argument that France was still a responsible nation which can be trusted by the world community to run the Antarctic?

The Antarctic Consultative Parties try to reassure the world that they know best how to run the Antarctic and are doing a better job than (say) the UN ever could. French nuclear testing would erode that claim.

It seems incredible that France would behave so stupidly. Indeed, its 1989 decision to oppose the mining treaty would seem to suggest an increased French concern for protecting Antarctica.

Australian journalist Padraic McGuinness (who has not addressed the speculation over nuclear testing on the Kerguelens) has warned that France's new-found interest in Antarctica may not necessarily please environmentalists:

> There is to be constructed a new Antarctic winter research base in the interior of the continent, the third after those of the United States (Amundsen-Scott) and the Soviet Union (Vostok). This will cost some 35 million francs (about A$8 million).
>
> At the same time, French polar research is being totally re-organised and stepped up, with three elements – an interministerial committee under the Prime Minister, a polar environmental committee, and an operational arm, the Institute for Polar Research. This will be concerned with the Antarctic and sub-Antarctic French interests, including the Kerguelen Islands, Crozet Island and Amsterdam Island.
>
> In part this re-organisation is for administrative reasons. The new set-up derives from consultations early last year [1989] between various governmental agencies – the ministries for overseas departments and territories (DOM-TOM), research, finance, foreign affairs, sea and transport, environment, and defence. But the scientists actually working in the polar bases were invited neither to participate in the discussions nor to give their advice.
>
> A report was given to the Minister for DOM-TOM and the Minister for Research and Technology on November 24 last. The scientists in all the relevant disciplines who had been excluded from all this met on December 16 and protested against the

> re-organisation; they met again on February 3 with the Minister for Research and the various bureaucrats concerned and reiterated their complaints, but agreed to work in the polar regions.
>
> The point of all this is that it is very clear that the driving force of French Antarctic research initiatives is neither scientific nor environmental, but derives from external affairs and defence considerations.[17]

McGuinness concluded:

> It has to be said that pointing all this out does not imply any hostility to France. But it does involve a recognition of the fact that France's strategic and economic interests may not in the future coincide with Australia's.
>
> We have seen how stupidly a French government, that of Mr Jacques Chirac, can act in our immediate region when it chooses to do so for domestic political reasons or on account of a nostalgia for empire. A future French government could act similarly in Antarctica.

Antarctic waste disposal and oil spills

A third area of increasing international scrutiny is the disposal of waste in Antarctica. Can the Antarctic expeditions be trusted to clean up their own mess?

The US base at McMurdo Sound, run by the US National Science Foundation, is an example of this problem:

> Those who have been there say it looks like a dusty mining camp in the Old West – a jumble of buildings and dirt roads on dark, barren volcanic ash.
>
> A pipe leading from the station to the shore pours raw sewage into McMurdo Sound.
>
> Not far away, abandoned trucks, rusty barrels, old tyres and other junk cascade down an embankment into the water.
>
> Smoke rises from open rubbish burning at the base dump. Waste oil oozes from dozens of deteriorating drums and dark stains advance across the barren ground.
>
> These are the conditions that photographers for Greenpeace, the international environmental organisation, have documented since the group assigned itself the role of watchdog for the Antarctic two years ago.
>
> 'They're doing things down there in an essentially pure environment you wouldn't be able to do in the US,' said Paul

> Bogart, who heads the Greenpeace Antarctic Campaign in the United States.
>
> US laws prohibit open burning of rubbish or dumping junk into coastal waters.[18]

The problem was well publicized in January 1989 when an Argentinian ship, the *Bahia Paraiso*, hit rocks in the Bismarck Strait on the northern tip of Antarctica – the worst environmental disaster yet to occur in Antarctica. The ship was trying to negotiate a strait clearly marked on US–UK charts as containing dangerous ledges and pinnacles. In short, the disaster could have been avoided. Scientists at a nearby US research base said fuel oil had washed ashore around the vessel, putting thousands of penguins and other seabirds at risk. Sixty per cent of the skua chicks, a marine bird, died or disappeared. The precise number of chicks affected was not known. At least six penguins had been found dead. The Chilean navy placed a barrier of floating tubes around the vessel to contain the spread of oil. But officials at the Chilean Antarctic Institute said patches of oil had been carried by the winds and tide up to 80 kilometres from the site of the wreck.[19]

The year 1989 was a bumper one for oil spills. Senator John Button (representing in the Senate the Australian Minister for Foreign Affairs and Trade) told the Senate in October 1989 that the Australian Government was worried about the reported oil spills from US stations:

> The Government has been concerned to hear reports of two recent fuel-oil spills at a United States station in Antarctica. The first, said to be up to 200,000 litres, was discovered in late September at the United States South Pole base. The second, reportedly involving up to 225,000 litres of jet fuel, diesel oil and petrol, was discovered on October 11 at Williamsfield Air Base on McMurdo Sound near the coast. In response to our inquiries United States agencies have limited comment to acknowledgment of the leaks, their causes and rectification and, in the case of McMurdo, the fact that clean-up of the area has begun. The United States Science Foundation maintains that when the clean-up is completed, environmental impact is expected to be minimal.[20]

The Minister also mentioned that besides these US problems there were also in 1989 four oil spills from other nations (one of which presumably was the *Bahia Paraiso*).

During 1990 the studying of the effects of oil spills in Antarctica became a growing area of scientific research. Australia contributed to both the research and the problem. Australia's first serious fuel spill took place in June 1990. The spill took place at the Casey base. To examine the tank that leaked 90,000 litres of Special Antarctic Blend fuel, members of the expedition at Casey had to use a bulldozer to cut away metres of ice from around it. (They first had to dig out the bulldozer in mid-winter dark!) Biodegradation – the natural breakdown of hydrocarbons in the oil – has an icy straitjacket in Antarctica. It is speculated that the process is 100 times slower than in temperate areas. Casey workers finally found the cause: the oil escaped when the tank settled on its base, cracking a rigidly iced elbow joint. The fuel was recovered but research was required to detect the environmental damage of the spill.[21]

A lawyer from the Washington DC-based Environmental Defense Fund, Bruce Manheim, was returning from a visit to Antarctica at the time of the *Bahia Paraiso* problem. He said the spill of the fuel oil was just one example of disregard for the Antarctic environment. He brought back what he said was the first film of a large, open garbage dump near the Argentinian Esperanza base. He said in order to reach a nearby breeding ground, penguins were forced to waddle across broken glass and scrap metal. The US McMurdo base, the region's largest outpost with a staff of up to 1,500 people, had a large open landfill. Mr Manheim said the waters surrounding the US base contained greater concentrations of cancer-causing PCB chemicals and heavy metals 'than virtually all the waterways in the United States. These toxins are now found in the tissues of Antarctic penguins and seals'. Other problems include a French airstrip cutting a penguin nesting ground from the sea; dynamite blasting near an Indian research station: establishment of bases by Uruguay and China in what was to be a specially protected area; and use by Chile of a lake as a garbage dump.[22]

This problem is bound to get worse. During the summer months, some research bases become, in effect, small towns. Antarctica now has up to 69 bases operating. Additionally, about 3,000 tourists visit Antarctica each year. There is no code of conduct for the tour operators. Once again, the waste problem is a poor reflection of the ability of the Consultative Parties to safeguard Antarctica's environment.

Notes

1. Abdul S. Minty, *Apartheid's Threat to World Peace*, New York, UN Centre Against *Apartheid* 1977, pp. 23-4.
2. Richard Woolcott, 'Antarctic Politics and Marine Resources: Critical Choices for the 1980s', *Australian Foreign Affairs Record*, Canberra, June 1984, p. 599.
3. See: Zdenek Cervenka and Barbara Rogers, *The Nuclear Axis: Secret Collaboration Between West Germany and South Africa*, New York: Times Books, 1978.

4. Frank Barnaby, *The Militarization of South Africa*, New York, UN Centre Against *Apartheid* 1978, pp. 4-5.

5. *South Africa's Plan and Capability in the Nuclear Field*, New York, UN Department of Political and Security Council Affairs, 1981, pp. 31-3.

6. Nan Rice, *Marion Island – A New Disaster?*, Fish Hoek, Dolphin Action and Protection Group, 1987.

7. 'Gain for Marion', *Cape Times*, 14 May 1987.

8. 'The French Airstrip – A Breach of Antarctica Treaty Rules?', *Antarctica Briefing No 9*, Washington DC: The Antarctica Project and Greenpeace International, July 1986, pp. 1-2.

9. These views were set out in more detail in: Keith D Suter, 'Kerguelen: A French Mystery', *Island Magazine* (Hobart) Summer 1987, pp. 8-14.

10. Mark Delmage, 'Mururoa Move?', *Peace Magazine* (Melbourne) April 1987, p.9.

11. 'French May Move to Test Site off WA', *The Australian*, 1 December 1984.

12. People for Nuclear Disarmament, *Nuclear Weapon Testing in the Indian/Southern Oceans*, Hobart (no date), p. 2.

13. 'French Nuclear Testing and the Sinking of the *Southern Raider*' *Australian Foreign Affairs Record*, Canberra, November 1986, p. 1038.

14. Ibid.

15. Brenton R. Groves, 'Nuclear Tests by France' (letter) *The Canberra Times*, 5 April 1982.

16. F.M. Auburn, *Antarctic Law and Politics* London, Hurst, 1982, p. 222.

17. Padraic McGuinness 'France Skates on Thin Ice', *The Australian*, 21 February 1990.

18. 'Watchdog for the Antarctic', *The West Australian*, 1 October 1988.

19. 'Navy Divers Stop Wreck's Oil Leak', *The West Australian*, 11 February 1989.

20. 'Antarctica: Oil Spills', *The Monthly Record* (Canberra) October 1989, p. 594.

21. 'Antarctica Fuel Spill Cause is Identified', *The Sydney Morning Herald*, 30 June 1990.

22. 'Antarctic Areas "Becoming Dump" ', *The West Australian*, 8 February 1989.

7. The Greening of Antarctica

This chapter deals with Antartica's emergence as an important issue via the wider consideration of protecting the global environment. In other words, Antarctica itself was not the focal point but ideas about protecting the global environment have had to include Antarctica.

International commissions

The 1980s have been the decade of international commissions of enquiries. Distinguished political leaders have been recruited to serve on enquiries which have been independent of national governments and yet whose reports have all encouraged governments to take a broader, more internationally-minded approach to global problems: international problems require international solutions. Former West German Chancellor Willy Brandt chaired two enquiries into aid for Third World Countries,[1] former Swedish Prime Minister, the late Olof Palme, chaired an enquiry into disarmament,[2] Willy Brandt and Michael Manley, former Prime Minister of Jamaica, convened an enquiry to look further at the two Brandt Reports[3] and Prince Sadruddin Aga Khan, former UN High Commissioner for Refugees, and Crown Prince Hassan bin Talal of Jordan chaired an enquiry into humanitarian issues.[4]

The success of these reports (all have been best sellers) is due partly to the unusual characteristics of the 'commission' approach. Since members of the independent commissions serve in an individual capacity, they do not have the same constraints as governments. Commissions can directly address public opinion, the media, social movements, political parties, and similar groups, and can structure and shape the issues in a way that makes substantive sense, unrestrained by the diplomatic agenda and unhampered by the formalities and procedures that govern the diplomatic process. Commissions can generate and stimulate transnational and international processes of discussion that can supplement and complement the intergovernmental organizations. Commissions can mobilize and interact with top scientists and experts without the formal constraints that often apply to national governments. Commissions can examine the linkages and inter-relationships that obtain across issue-areas without the jurisdictional delimitations that are imposed on national governments and international institutions.

All these reports recommended international co-operation. They were all, in effect, replying to the prevailing philosophy of the 'New Right' in Western governments which stressed the cult of the rugged individual, making a virtue out of national and personal selfishness, doing unto others before they have a chance to do it unto you, reducing governmental co-operation through the UN, and expecting the free market system to solve problems.

This philosophy of co-operation may be seen in two environmental studies which, among other things, called for changes to the Antarctic Treaty System.

World Conservation Strategy

The World Conservation Strategy was devised in 1980 by the International Union for Conservation of Nature and Natural Resources (IUCN), an organization containing 450 governmental agencies and non-governmental organizations in 100 nations, the UN Environment Programme and the World Wildlife Fund (one of the world's largest environmental non-governmental organizations).[5] The Strategy represented the first time that these organizations had jointly produced an expert study to show how conservation can contribute to the development objectives of governments, industry, commerce, organized labour and the professions. The Strategy argued that economic development could be viewed as a major means of achieving conservation – instead of being viewed as an obstruction to it. It emphasized the need for humankind to be more humble in its use of nature and to recognize the adverse impact of rapid economic growth on the environment. Only by co-operating with nature could humankind expect to survive.

The goal of the World Conservation Strategy was the 'integration of conservation and development to ensure that the modifications caused by both in the environment, secure the survival and well-being of all people'. This was broken down into three main objectives: (i) To maintain essential ecological processes and life support systems on which human survival depends – for example, to cleanse waters, regenerate soils. (ii) To preserve genetic diversity – that is, to protect species of wild life both animal and vegetable and use them to enhance diversity, and improve cultivated plants and trees, domestic animals, fish and micro-organisms. (iii) To ensure the sustainable utilization of species and ecosystems – that is, keeping forests, fishing grounds, arable and grazing lands renewed for supplying food and other resources to support rural communities and industries.

The Strategy set out a framework for governments to develop national conservation strategies to operate within the overall World Conservation Strategy, and it identified priorities for international co-operation.

One area for international co-operation was the 'global commons':

> A commons is a tract of land or water owned or used jointly by the members of a community. The global commons includes those parts of the earth's surface beyond national jurisdictions – notably the open ocean and the living resources found there – or held in common – notably the atmosphere. The only land mass that may be regarded as part of the global commons is Antarctica, although several countries have claimed parts of it (the claims are currently frozen under the Antarctic Treaty).

The Strategy then looked in detail at the 'open ocean and its living resources', the atmosphere and climate, and Antarctica and the Southern Ocean.

The Strategy document summarized Antarctica's geography and its impact on global climate patterns. It noted the need for caution in harvesting krill because of krill's importance for the Antarctic food chain. This should be a key component for the living resources treaty which was being finalized at that time (see Chapter 3 above). It recommended that 'an independent observer system should be provided for in such regulations'. This recommendation was not included in the Convention on the Conservation of Marine Living Resources. For reasons elaborated in this book, such a system would be of great use.

The Strategy document noted the lack of agreed data on, for example, the maximum amount of krill that can be harvested without endangering its future (and thereby the rest of the Antarctic food chain). It recommended an 'International Decade of Southern Ocean Research, focussing particularly on ecological processes'. This recommendation has not been followed up.

Finally, the strategy document noted the interest in oil exploration. It recommended that 'the feasibility of oil exploration and exploitation in particular should be approached with the utmost caution'. This recommendation has, in effect, been adopted – but this is due mainly to the current comparatively low price of oil. As Chapter 4 has explained, some Third World countries and non-governmental organizations have suspicions that eventual oil exploitation is a possibility.

At first sight, these recommendations appear mild and unadventurous. But three points need to be remembered. First, it is impressive that Antarctica is included in this document as a 'priority for international action'. After all, the Strategy was devised in 1980 – two years before Antarctica became a fixture on the UN General Assembly's agenda. Second, Antarctica is alone in the context of 'global commons'. As Chapter 5 showed, the Antarctic Treaty Consultative Parties are rather nervous about references to 'commons' in regard to their territory. Third, a decade on, all three recommendations are still relevant and two remain to be implemented.

Our common future

In 1984 the United Nations set up an independent commission of world leaders to consider ways in which nations could co-operate to attack the twin problems of poverty and the destruction of the environment. The World Commission on Environment and Development (WCED)[6] was chaired by Gro Harlem Brundtland, Norway's first woman prime minister.

In carrying out its mandate, WCED decided that it had three main tasks to fulfil: to examine all the important world issues of environment and development and then formulate realistic proposals for dealing with them; to strengthen existing international co-operation and to suggest new and more effective means of dealing with such issues; and to persuade people to take the problems seriously.

At the beginning of its work, the Commission considered two alternative ways of looking at the critical issues of environment and development. The first strategy, the one most used, involved looking at the issues very much in isolation from each other. For example, it could look at carbon dioxide build-up in the atmosphere, acid rain and lead pollution, all as separate issues. The alternative approach, and the one that WCED decided to adopt, was a more 'holistic' one that would examine the underlying connections between different critical issues.

The economies of many Third World countries are caught in a multiple bind. They must improve housing and roads, and develop industries and agriculture, at a pace that industrialized nations never had to contemplate. They have to export more of their resources to earn the external capital for such development. But with prices of commodities falling, and protectionism in the West increasing, these developing countries find themselves in a situation of rapidly declining development.

Moreover, in pursuit of increasingly elusive foreign exchange – needed not only for development, but also to service foreign debt – Third World societies often destroy their own environment; not because they are ignorant, but because they need to survive. They over-cultivate already thin soils, they deplete grasslands by over-grazing, and they demolish forests for firewood which they use as fuel to cook their food. The consequences are: deforestation, desertification, and, eventually, more destitution. Poverty becomes both a cause and an effect of environmental degradation.

Environment has come to imply a technical afterthought, exemplified by the anti-pollution device added on to existing machinery. In the same way, it is often a political afterthought too. A major problem such as desertification can be tackled after it has happened but political forethought would see it tackled at its source in agricultural and rural settlement policy. Such shifts in policy will be difficult to define and promote.

The Brundtland Report, *Our Common Future*, focuses on population, food security, the loss of species and genetic resources, energy, industry, and human settlements. It points out that these issues cannot be treated in isolation from one another.

One chapter of *Our Common Future* deals with 'Managing the Commons'; the 'commons' consist of the ocean, outer space and Antarctica. This report is many times larger than the World Conservation Strategy and so, as may be expected, it devotes many more pages to Antarctica.

The report recalls that Antarctica is now on the UN General Assembly's agenda.

> Under the combined pressures of economic, technological, environmental, and other trends, there are new initiatives to establish a regime for minerals exploitation. New questions about equitable management are presenting challenges that may reshape the political context of the continent within the next decade. During the forthcoming period of change, the challenge is to ensure that Antarctica is managed in the interests of all humankind, in a manner that conserves its unique environment, preserves its value for scientific research, and retains its character as a demilitarized non-nuclear zone of peace.

The report notes the recent political controversy over how Antarctica's destiny should be handled.

> The Commission does not propose to adjudicate the status of Antarctica, but it sees it as essential that the continent be managed and protected in a responsible manner that takes into account the common interests at stake. It notes also that the legal and management regimes are in the midst of a process of change leading to wider participation.

The report would like to see a 'guarding of present achievements':

> Although further change in the management status of Antarctica is inevitable, it is essential that such change not jeopardize the achievements of the Treaty System in the areas of peace, science, conservation, and environment. Antarctica has been an agreed zone of peace for nearly 30 years, free of all military activities, nuclear tests, and radioactive wastes. This is a foundation on which humanity must build.

As with the World Conservation Strategy, the report is worried about the potential impact of mineral exploitation.

> Given the absence of technologies tested in the ultimate extremities of Antarctic conditions, the lack of agreement on procedures to assess and take account of the impacts of any development, and the sparse data base, it could take a generation or more of dedicated research and technological development to ensure that minerals exploitation would not destroy the Antarctic's fragile ecosystem and its place in global environmental processes. Thus it is important that no minerals activity takes place until these conditions have changed, and then only in consonance with a regime that guarantees implementation of the most stringent standards needed to protect the continent's environment and share the proceeds equitably.

In terms of specific recommendations, this report is rather disappointing. It calls for the 'evolution' of the Antarctic Treaty System.

> In the years ahead, activities in Antarctica will expand in kind and scale, as will the numbers of participants in such activities. Further efforts must be made to ensure effective management of those activities and an orderly expansion of participation in such management. A variety of options are being discussed by the international community. More effective management, including expanded participation, could evolve gradually through the existing Treaty System. But given the extent of probable change and the lure of mineral wealth, however remote, such an approach could be too slow to retain political support. Another is that the above goals might be reached through the negotiation of an entirely new system. However, neither of these approaches would be free of difficulty. Yet another alternative would be to intensify efforts to make the Treaty System more universal, more open, and responsive to expressions of concrete and legitimate concern and interest in Antarctica.

This is a rather confusing range of ideas. In fairness to the Commission, I suppose it could be said that the Commission reflects its particular era: there is no agreement at all at this time (as Chapter 5 has shown) as to how Antarctica ought to be managed.

Thus the report's final paragraphs focus on process rather than substance: the need to establish 'a means for more effective communication':

> To focus on longer-term strategies to preserve and build on the achievement of the existing Treaty system, nations must create the means to foster dialogue among politicians, scientists, environmentalists, and industries from countries within and outside it. A good place to start would be the development of closer working relationships between the parties to Antarctic regimes and the international organizations within and outside the UN system that have responsibilities for science and technology, conservation, and environmental management.
>
> National policy processes could also be structured to provide for dialogue with concerned industries, public interest organisations, and expert advisors, perhaps through an Antarctic advisory committee. The US Government has been in the forefront of those countries appointing industry and public interest advisors to its delegations to Consultative Parties meetings. Australia, Denmark, and New Zealand have more recently followed suit.

The gap in the ozone layer

> Greenhouse gases and ozone depletion have brought Antarctic research into high public profile and created, rather suddenly, a transnational perspective on human fate. Interest in Antarctic research in these areas is apt to grow rapidly for one reason: adverse effects on global climate are likely to appear in Antarctica first, because of the central role the continent plays in the Earth's weather and because of the pristine nature of its physical environment. Antarctica serves, then, as an early-warning station and, with the information in its ice cores, as a sort of archive for the atmosphere.[7]

Ozone is a colourless, odourless gas that resides largely in the stratosphere, a region of the atmosphere which extends from about eight kilometres at the Poles and 17 kilometres at the Equator to about 50 kilometres above the earth's surface. The gas absorbs harmful ultraviolet radioaction from the sun and reduces the radiation to safe levels at the earth's surface. It is, then, one of the most vital components of the earth's life support systems. Now, however, there is concern about the layer's future because of human activities, such as the development of chlorofluorocarbons (CFCs).

Ironically, CFCs were once considered ideal industrial chemicals because they are non-toxic, non-flammable and highly stable. Since their invention in 1928, they have been used to cool refrigerators and air-conditioners, propel aerosol sprays, clean computer chips, and to make foam rubber and polystyrene containers common in food packaging.

The first intimations of an environmental problem with chlorofluorocarbons came in the late 1960s, when the British chemist James Lovelock developed a device that detected airborne CFCs. To his surprise, one of his devices on a ship sailing in the Southern Hemisphere found traces of CFCs, far from any major release point. Once released into the atmosphere, CFCs apparently never disappear. They seem instead to spread indefinitely, mingling with other molecules in the air until they are dispersed over the globe.

In 1974 two US chemists, Sherwood Rowland and Mario Molina, first warned that the ozone layer was threatened by CFCs. They reported that CFCs were being added to the atmosphere in steadily increasing amounts. At that time annual production of the compounds in the United States alone was over 380 million kilograms. CFCs, according to Rowland and Molina's theory, remain intact in the atmosphere for 50 to 100 years, slowly drifting into the stratosphere. There, intense ultraviolet radiation severs the compounds' chemical bonds, releasing chlorine atoms in a catalytic chain reaction that threatens stratospheric ozone. Rowland and Molina calculated that for every chlorine atom released, 100,000 molecules of ozone are removed from the atmosphere. This theory was controversial among atmospheric scientists, and public reaction to the research was muted.

A reduction in the ozone layer will cause more skin cancers and may trigger climate changes. For example, a one per cent decline in the ozone layer could lead to a four to six per cent increase in the incidence of some skin cancers. By the middle of next century the incidence of skin cancers could increase by 20 to 48 per cent as a consequence of past and current chlorofluorocarbon production.

Antarctica has received attention in this context because the first evidence of ozone depletion was located over Antarctica. The hole over Antarctica came as a shock. It had been detected in 1985 but the ozone readings were so low they were discarded as a product of faulty equipment. Corroboration by the UK and US confirmed a spring fall by 40 per cent in ozone above Antarctica.

Confirmation of the ozone hole was not necessarily an indictment of CFCs. The stratosphere over Antarctica is unlike any other section of the earth's atmosphere; it is dominated by the polar vortex, a tight whirlpool of winds. When the vortex breaks up every November, the region is replenished by ozone-rich air from other parts of the stratosphere. Some scientists speculated that these atmospheric winds or other cyclical weather effects could explain the hole. In August 1986 a team of scientists went to Antarctica and conducted the first thorough analysis of the continent's stratosphere. The expedition's experiments strongly suggested that CFCs played a role in the destruction of stratospheric ozone.

Discussions among the industrialized nations and the United Nations Environment Programme (UNEP), following the expedition, paved the way for the 1987 Montreal Protocol on Substances that Deplete the Ozone Layer. US negotiators at Montreal initially sought an immediate freeze on CFC production, followed by a 95 per cent reduction to be phased in over a number of years. Instead, the treaty called for a freeze on production at 1986 levels beginning in 1990, followed by a 50 per cent reduction in CFC emissions by the end of the century. Many developing nations, in need of inexpensive refrigeration, were exempt from these limits for ten years.

In October 1988, UNEP convened an international conference of scientists to review recent research on the ozone layer. They considered the results of recent research on ozone in the atmosphere where observations confirmed smaller amounts than were earlier calculated. They gave particular attention to the seasonally occurring Antarctic ozone hole and its global implications, and to similar but smaller holes observed in Arctic areas.

Even if the regulatory measures contained in the Montreal Protocol were applied, the predicted atmospheric burden of chlorine would result in the Antarctic hole remaining forever. Stronger regulatory measures (close to a complete phase-out of emissions) were necessary.

A ministerial meeting was held in London in June 1990, to consider amendments to the Montreal Protocol. The principal issues facing the meeting were: setting dates for the phase-out of use of the main ozone-depleting substances, transfer of relevant technology and the creation of a fund whose main purpose would be to facilitate observance of the Montreal Protocol by developing countries.

Considerable success was achieved in London in moving up the dates fixed in the Montreal Protocol for the phasing out of chlorofluorocarbons. The Protocol has called for 50 per cent phasing out of CFCs by the year 2000. This has been accelerated to 1995; with 85 per cent by 1997 and 100 per cent by 2000. Thirteen countries issued a statement at the meeting declaring their firm intention to phase out CFCs no later than 1997. Other major ozone-depleting substances will be subject to similar accelerated phasing out, although some concern has been expressed at the failure of the meeting to apply control measures to a further group of transitional substances. The latter will be reconsidered at the next review of the Montreal Protocol in 1992.

The other major outcome of the London meeting was agreement to create a US$160 million fund within the UN system over the period 1991–93. The fund's initial capital may increase to $240 million if India and China become parties to the Montreal Protocol, as they indicated they might. Developing countries party to the Protocol will be able to use this fund to redesign or develop technology to minimize use of ozone-depleting substances.

The resolution of the technology transfer issue was less satisfactory, and questions of implementation remain to be resolved. The main point at issue was one of approach. Developing countries, led by India, argued that they could comply with control measures established by the Montreal Protocol only if relevant technology was transferred to them. The developed countries, in general, argued the reverse, that technology could be transferred only to the extent practicable to developing countries complying with the control measures. The final language reached was a diplomatic compromise, but clearly, further work was needed before the issue could be resolved to the satisfaction of all.[8]

The significance of the ozone problem for this book is partly that it draws international attention to Antarctica. More generally, the ozone problem is a reminder of the transnational nature of environmental problems and the need for imagination in devising new ways of coping with those problems. Governments at present are more receptive to new ideas and so there is a window of opportunity for new ideas on Antarctica.

The greenhouse effect

The world first heard about the 'greenhouse effect' over half a century ago.[9] In 1938 an air pollution specialist in the UK, G.S. Callendar, suggested that fossil-fuel burning must be increasing the carbon dioxide in the atmosphere. Carbon dioxide lets sunlight through, but slows the rate at which the consequent heat radiates back into space, hence the name. Since then, other gases have turned out to be important in this effect: they include methane, ozone, nitrous oxide, water vapour and CFCs.

The earth is surrounded by a blanket of air: the atmosphere. Solar radiation passes through the earth's atmosphere and warms the land and sea. Thirty per cent of the heat radiated back from the earth's surface (in the form of infra-red radiation) is absorbed by water vapour and carbon dioxide in the atmosphere, and the remaining 70 per cent eventually ends up in space. Because the air is warm, it too radiates heat and some of this heat goes back towards the ground, warming it up further. Provided the amount of carbon dioxide and water vapour in the atmosphere remain constant, and the sun gives off the same amount of heat, an equilibrium is maintained. This is the greenhouse effect. It is a natural phenomenon which keeps the earth liveable.

Because the greenhouse gases (carbon dioxide, water vapour, and a few others) not only absorb, but also emit radiation, if we increase the amount of these gases, the atmosphere will get warmer, and the greenhouse effect will intensify. This is what is now happening. The amount of carbon dioxide in the atmosphere has been steadily increasing since the beginning of industrialization – from the burning of coal and other fossil fuels. The developing world has contributed with the massive burning-off of tropical rainforests which, when alive, provide biological storage of carbon dioxide. As the earth's temperature rises, more water vapour is produced – which prevents more heat escaping and causes a further rise in temperature.

Human activities are releasing several other gases that absorb infra-red radiation at the wave-lengths at which it normally can escape from the earth's atmosphere. Nitrous oxide is produced from car exhausts and factory chimneys, and methane is produced by bacterial activity in paddy fields, from cows, landfills, and the release of natural gas from commercial gas and oil fields and open-cut mines.

By the year 2030, the atmospheric temperature is expected to rise by 1.5–4.5 degrees Celsius, and sea levels are expected to rise by at least half a metre as a result of thermal expansion and the melting of ice from the Antarctic ice cap. No one is exactly sure of the extent of these changes as there are not yet enough data to make accurate predictions.

It is beyond this book's scope to comment in detail upon what should be done about the greenhouse effect (and the hole in the ozone layer). They are important in Antarctica's context – and so should be mentioned here – for three reasons.

First, some of the research into these problems has been done in Antarctica. For example, Stephen Schneider (a US specialist in climatology), in reviewing the crisis of global warming – 'which should be unmistakable within a decade or two' – commented upon the value of research into Antarctica's ice core. A segment of a two-kilometre core drilled from the Antarctic ice sheet at the Soviet Union's Vostok Station contained trapped bubbles of ancient air. Analysis of the bubbles and of the ratio of hydrogen isotopes in the ice, which varied with local temperature, enabled scientists to reconstruct a 160,000-year record of trace gases and temperatures.[10] This provided an important base against which to compare recent climate change. In other words, Antarctica will occupy more public attention not just because of political pressure at the UN, but also because it will receive more mentions in the context of scientific research into the greenhouse effect and the hole in the ozone layer.

Second, Antarctica will also receive more attention because of speculation over the consequences of the greenhouse effect and the hole in the ozone layer. Antarctica is in darkness for six months a year but it still receives as much sunlight during its six-month-long summer day as the equatorial regions. Yet all that light creates virtually no surface heat, because the ice and snow reflect most of it back into space.

But as greenhouse gases trap more heat in the earth's atmosphere and temperatures rise, they are expected to climb fastest in Antarctica, because any melting at the edge of the ice shelf will change the reflectivity of the surface exposed to sunlight. More of the sun's rays will be absorbed by the larger ocean area, in turn magnifying the local warming trend, melting more ice at the margin and absorbing yet more radiation from the sun. As a result of the rising temperature the world's oceans would begin to rise, and dangerously so if the massive West Antarctic ice shelf eventually broke up and toppled into the sea. If that happened, the sea could rise by nearly five metres, submerging the world's low-lying cities and even entire nations (the Maldive Islands are less than four metres in height).

Because temperatures at the poles increase far more rapidly than at the equator, the very pattern of the world weather system might then change. In the past, distribution of heat from the equator to the poles and cold from the poles to the equator has been accomplished through an essentially constant pattern of wind and ocean currents. But if temperatures rise disproportionately in Antarctica, the equilibrium could change, creating new patterns that would redistribute less cold and more heat. In short, as speculation about environmental change gathers momentum, so will extra public attention fall upon Antarctica.

Finally, the greenhouse effect and the gap in the ozone layer are rare environmental crises: very few (if any) environmental crises have ever shot on to the political agenda with the speed these have done. By contrast, for example, Britain's 1956 Clean Air Act had been preceded by over a century of controversy about the health implications of smog. London was internationally notorious for its 'pea-soup' fogs and they were immortalized in Hollywood movies as backdrops for sinister scenes.

Governments have been anxious to demonstrate to their citizens that they are addressing the current crises. In March 1989, for example, two high level governmental conferences took place within a few days of each other (London and The Hague) to address these issues. Mrs Thatcher, a new convert to the environmental cause, hosted the London Conference.

The Hague Conference (hosted by The Netherlands, Norway and France) overshadowed the London Conference in terms of media coverage. 'The Declaration of The Hague', which was published in many countries at government expense, said that there had to be '. . . a new approach, through the development of new principles of international law including new and more effective decision-making and enforcement mechanisms.' The 24 signatory nations, drawn from a cross-section of the world's political systems, identified the following principles to be adopted in tackling the greenhouse effect:

(a) The principle of developing, within the framework of the United Nations, new institutional authority, either by strengthening existing institutions or by creating a new institution, which, in the context of the preservation of the earth's atmosphere, shall be responsible for combating any further global warming of the atmosphere and shall involve such decision-making procedures as may be effective even if, on occasion, unanimous agreement has not been achieved;

(b) The principle that this institutional authority undertake or commission the necessary studies, be granted appropriate information upon request, ensure the circulation and exchange of scientific and technological information – including facilitation of access to the technology needed – develop instruments and define standards to enhance or guarantee the protection of the atmosphere and monitor compliance herewith;

(c) The principle of appropriate measures to promote the effective implementation of and compliance with the decisions of the new institutional authority, decisions which will be subject to control by the International Court of Justice;

(d) The principle that countries to which decisions taken to protect the atmosphere shall prove to be an abnormal or special burden, in view, inter alia, of the level of their development and actual responsibility for the deterioration of the atmosphere, shall receive fair and equitable assistance to compensate them for bearing such burden. To this end mechanisms will have to be developed;

(e) The negotiation of the necessary legal instruments to provide an effective and coherent foundation, institutionally and financially, for the aforementioned principles.[11]

This is a very important document. It recognizes the interdependence of the globe; the need for common action to address common problems;

it sees governments (rather than the free market system) as the prime movers in tackling this problem, and the need for new approaches to international law. Government declarations are not, of course, the same as international treaties. But there is at least some evidence here of increased governmental awareness of the need to protect the environment. Additionally, the greenhouse and ozone layer problems are long-term issues that governments are being forced to think about today. Most long-term issues in politics are ignored by politicians, whose concentration span rarely exceeds the horizon of the next election. These problems are, then, introducing politicians to the need to take a long-term perspective in policy formulation.

The developments since the Hague Conference go beyond this book's scope. But the important link between that Conference and Antarctica comes from both the new sense of urgency and the legal vision represented by the Conference document. These are just the qualities required in working towards a new regime for Antarctica. Much, therefore, depends upon this unique opportunity being followed up.

Notes

1. Report of the Independent Commission on International Development Issues, *North–South: A Programme for Survival*, London, Pan, 1980; the Brandt Commission, *Common Crisis – North–South: Co-operation for World Recovery*, London, Pan, 1983.
2. Independent Commission on Disarmament and Security Issues, *Common Security: A Programme for Disarmament*, London, Pan, 1982.
3. Willy Brandt and Michael Manley, *Global Challenge – from Crisis to Co-operation: Breaking the North–South Stalemate*, London, Pan, 1985.
4. Independent Commission on International Humanitarian Issues, *Winning the Human Race?* London, Zed Books, 1988.
5. *World Conservation Strategy*, Gland, Switzerland, IUCN, 1980. Also see Robert Allen, *How to Save the World: Strategy for Conservation*, London, Kegan Paul, 1980.
6. World Commission on Environment and Development, *Our Common Future*, London, Oxford University Press, 1987.
7. Barry Lopez, 'The Cold, Clear View from the South Pole' *Dialogue* (Washington DC) No. 1, 1990, p. 31.
8. 'London Meeting on the Ozone Layer', *Backgrounder*, Department of Foreign Affairs and Trade, Canberra, 13 July 1900, p. 3.
9. Books on the 'greenhouse effect' are now a minor publishing industry. Among recent books are: Ken Coghill (Editor) *Greenhouse: What's To Be Done?* Sydney, Pluto, 1990; John Daly *The Greenhouse Trap*, Sydney, Bantam, 1989; David Everest *The Greenhouse Effect: Issues for Policy*

Makers, London, Royal Institute of International Affairs, 1988; J. Falk *The Greenhouse Challenge: What Is To Be Done?*, Melbourne, Penguin, 1989; Michael Grubb *The Greenhouse Effect: Negotiating Targets*, London, Royal Institute of International Affairs, 1989; Fred Pearce *Turning Up the Heat*, Sydney, The Bodley Head, 1989; Stephen H. Schneider *Global Warming: Are We Entering the Greenhouse Century?* New York, Sierra Club Books, 1989.

10. Stephen H. Schneider, 'The Changing Climate', *Scientific American*, September 1989, p. 43

11. 'The Declaration of The Hague', reprinted in *Backgrounder*, Canberra, Department of Foreign Affairs and Trade, March 1989, p. 3.

8. Antarctica and the Fifth World

The Fifth World

Our Common Future notes the way in which some national delegations now contain non-governmental persons. This is one example of how environmental non-governmental organizations (NGOs) are able to participate in the discussions on Antarctica.

It is customary to describe the globe as being divided politically into First, Second and Third Worlds (referring to the Western developed countries, the Communist nations, and the vast majority of nations which try to avoid becoming entangled in the conflicts and squabbles of the other two Worlds). The London-based human rights organization, Minority Rights Group, has popularized the idea of a 'Fourth World' whose human rights are being violated, such as the Catholics in Northern Ireland, the blacks in the United States and the Aborigines in Australia.

There is now a 'Fifth World', made up of a growing number of people who have loyalties and interests that transcend political boundaries. They want to help people; people whom they may never meet. Such people are often at their most effective in non-governmental organizations (NGOs). It is these organizations – or the people who support them – that constitute the Fifth World.

NGOs are separate from the branches of governments, such as the public service, the armed forces and nationalized industries. Many do, however, receive governmental assistance, if only because (especially in local and national affairs) they provide services which the public service cannot perform, such as the Life Line telephone counselling service.

Churches are among the oldest of NGOs. Human rights organizations, such as Amnesty International and the International Commission of Jurists, are among the newest. Development groups, for example, Oxfam, Community Aid Abroad and World Vision, are NGOs that encourage and enable developed countries to assist developing countries. The anti-uranium movement is a recent addition to the Fifth World.

It is worth recalling that members of the Fifth World represent only a small proportion of the world's total population. The Fifth World is, however, growing rapidly and has an impact out of proportion to its size. Australia's conservation movement, for example, has 400,000 members (out of a total population in excess of 16 million) – but that membership exceeds the combined individual membership of all Australia's political parties. Indeed, it would seem that often people prefer to join NGOs and campaign on specific issues rather than join a political party and become bogged down in the petty politics of personalities and branch meetings. Skilled and dedicated people join NGOs and they can use their skills (as in public relations, for example) to great effect. Additionally, we should not underestimate the power of ideas. Two decades ago, 'pollution of the environment' received little attention in the mass media. When it did, it was usually in the limited context of dirty rivers and non-returnable bottles. Problems of the environment are now recognized as among the gravest in human history. People are now being encouraged to question even their life styles and cultures.

The neglected subject

> The study of world affairs has, from the beginning, been largely confined to the relationships between States. It has been concerned primarily with struggles for power, wars, the resulting problems and the efforts of States to obtain co-operation for peaceful and constructive ends. That study is still, therefore, in a stage of development similar to the study of political science in the United States some decades ago when students gave attention only to formally constituted branches of Government, not recognising, as they do now the important role of political parties and 'pressure groups', eg trade unions, chambers of commerce, churches.
>
> The time has come to realise that the unofficial side of world affairs has developed to the point where it should be given consideration as an important aspect of international life. In particular, the students of international organization should recognise the immense contribution of those organizations composed of unofficial groups in different countries which have come together to promote common interests through international action.[1]

That assessment, written in 1968, remains equally valid today. Indeed, the 'unofficial side of world affairs' has developed still further.

The startling growth of the Fifth World may be explained in three ways. First, improvements in communications technology are creating a new international spirit of oneness, both between humans, and between humans and the natural environment. Traditional divisions between nations and peoples (whether as a result of religion, geographical boundaries, or ideologies) no longer seem so significant. On the contrary, they are held partly responsible for most of today's problems, the arms race being a startling example. The Fifth World is more concerned with the common features that unite people (such as a pleasant life and a healthy environment), than with the old divisions that separated people.

Second, governments are shrinking in importance. This may seem a strange statement when so many people are worried about governmental use and abuse of power. But national governments are being submerged by new power groupings. International lawyers often use 1648 as the commencement date of the nation-state in modern history. The phenomenon of national governments has had a good run; but it is now ending. The creation of the UN in 1945 changed the role of national governments; the UN has absorbed many governmental responsibilities.

Also, international economic affairs are increasingly dominated by transnational corporations, the major ones are economically more powerful than most of the UN member-states. Even in the US, which hosts the head offices of many corporations, there is a fear that the 'national interest' and a 'corporation's interest' may no longer be identical. For example, writing in 1987 in readiness for the 1988 US Presidential election, Jeff Faux commented:

> There is still widespread distrust of large corporations and concentrations of private wealth: the abuses of corporate power have become common newspaper reading, and most Americans think that big corporations exert decisive influence on the US economy. Democrats can confront Reagan's laissez-faire ideology by presenting a clearly articulated analysis of large corporations' present transnational character, and arguing that such corporations should not be trusted with America's economic future – not because the people who run them are bad, but because their jobs do not require them to be concerned with the national interests. 'We're all doing it with a great deal of sadness,' said Robert Lutz of Ford about his company's overseas outsourcing of automobile manufacturing. 'But we have come to regard our primordial duty as the preservation of the company and the shareholders.' Such logic may be in Ford's best interest, but it is disastrous for the US economy.[2]

Meanwhile, many trade unions are gaining in power and their members can disrupt the national economy. The government's role in economic affairs, therefore, is declining from that of being the sole leader to being the referee between business and unions.

Meanwhile at the grass-roots, the Fifth World is gaining in influence simply because its members are losing faith in politicians. As one humourist has said: 'Whoever you vote for in an election, a politician always wins.' With governments unable to supply the long-term answers, the people are taking on that responsibility via the non-governmental organizations.

Finally, people in developed countries are keenly aware of the international crises and dubious of the traditional methods of settling them. They are tired of their governments responding to the immediate crises rather than looking ahead to the problems looming on the horizon and acting positively.

Michael O'Neill, of the University of San Francisco,[3] has examined the NGO situation in the US. The NGO sector, perhaps the biggest unknown success story in US history, employs more civilians than the federal and state governments combined, with a budget that exceeds all but seven nations in the world. Eighty million American adults and teenagers contributed US$150 billion of volunteer work within its organizations in 1987. He divides the non-profit arena into nine subsectors: religion, private education and research, health care, arts and culture, social services, advocacy and legal services, international assistance, foundations and corporate funders, and mutual benefit organizations. The US has 1,243,000 NGOs spread across these nine categories. The US is obviously the world leader – but similar studies in other nations would produce a similar pattern.

This book is concerned with only a minute fragment of the NGO sector: environmental advocacy groups – both in the US and elsewhere. But it is worth recalling the vast panorama of the NGO tapestry from which this small thread comes.

The term 'NGO' first entered international jargon in 1945 when they received formal recognition in the UN Charter (Article 71):

> The Economic and Social Council may make suitable arrangements for consultation with non-governmental organizations which are concerned with matters within its competence. Such arrangements may be made with international organizations and, where appropriate, with national organizations after consultation with the Member of the United Nations concerned.

Article 71, therefore, accorded formal recognition both to the co-operation rendered the world body by NGOs and to their future potential. The League of Nations, by contrast, provided for no such mechanism. The Economic and Social Council (ECOSOC) thereby obtains 'expert information or advice' from this consultative process and, at the same time, has access to important segments of public opinion.

An ECOSOC resolution of February 1950 spelled out liberal arrangements for consultation. Those arrangements have been exceedingly productive. In such economic fields as housing, building and planning, a number of specialized NGOs have provided invaluable assistance. The International Union of Architects, the International Union of Building Societies and Saving Associations, and the International Confederation of Free Trade Unions have been especially active in this respect. In the fields of transport, travel and tourism, too, various NGOs have contributed significantly. Once again the point is made that this chapter is concerned mainly with environmental NGOs and it is worth remembering that these are only a small part of the vast NGO array.

The work of NGOs at the international level

The first function of non-government organizations active at the international level is that of giving some status to the individual in international relations. For example, the only participants at international conferences who may speak, table resolutions and vote on resolutions are governmental delegates. Certain international entities, such as UN Specialized Agencies and regional inter-governmental organizations, like the Organization of American States, may also be given some, more limited, rights of participation (usually to speak, but not to vote). This is a reflection of international relations and international law; only governments may make international law and that law in the first instance usually applies only to governments.

Assuming that an individual is not a member of a government, the only way that he or she can hope to have any influence is by becoming a member of a non-governmental organization. An NGO may be invited to attend international conferences, where it will have certain rights, usually on a par with the inter-governmental bodies. In practical terms, this means the right to ask the chairperson to address the conference (permission is nearly always given), the right to listen to all public speeches, the right to receive all unclassified documents, the right to request that official facilities be provided for NGOs to have their own meetings. From an individual's point of view, the NGO is a vehicle for winning acceptance for an idea they have and which their own national government may not like; thus instead of wasting time in trying to get agreement from their own government to take the idea to an international forum (a diplomatic conference or the UN General Assembly) they can send it to an international NGO to take it direct to that forum.

The second function of these organizations is that of monitoring governmental behaviour at international conferences (as well as at meetings of UN bodies). On the one hand, these NGOs observe these gatherings and note what goes on. Since World War I, most conferences conduct most of their business in public, the summary records are published and journalists are inundated with press releases explaining what is going on in the conference. Not all of a conference's work is conducted in public. Because of the number represented at conferences, governments belong to one of five caucus groups (Western and others, East Europe, Africa, Asia and Latin America), plus occasional other groupings, such as the European Community. The value of this approach is that instead of a range of resolutions being tabled at the conference, it is usually possible to keep the number down to around five, thereby saving considerable money and time. These caucus meetings are nearly always held in private. But with the caucus work over, the resolutions tabled, the conference returns to public meetings and the organizations may observe what speeches are made and how delegates vote. Moreover, by being present in the conference precincts, observers from the organizations get to meet delegates either at the conference itself or at diplomatic receptions. In private conversations with delegates they are able to find out a great deal of what happens in the caucus groups.

The other side of this function is that of reporting back to the national branches of NGOs on what happened at the conferences. The fact that these NGOs exist is also a post-World War I development. As somebody commented soon after World War I, 'In the past, the people of most countries knew little, and probably cared less about the foreign policy of the government than about any other department of State life.'[4] World War I shattered this lethargy and resulted in a range of organizations being created, such as non-politically partisan academic bodies such as the national Institutes of International Affairs. For the average citizen interested in foreign issues, the media coverage is usually inadequate. NGOs, via their magazines and newsletters, provide better (and more specialized) coverage.

The final function of NGOs is that of being a source of views and research that transcends national boundaries. Despite the anachronism of many nations being dwarfed by transnational corporations and all nations being threatened by common problems, such as the depletion of the earth's non-renewable resources, the main actor in international relations continues to be the nation-state.

But the formal pre-eminence of nation-states is being reduced by NGOs. First, by drawing their membership from more than one nation, NGOs have a constituency very different from that of governments. They therefore represent a different perspective, a less parochial outlook, in international relations. As such they are able to put forward ideas which they believe are best for the world as a whole, rather than just one nation. Even governments recognize this by, for example, allowing NGOs to participate at conferences and meetings of UN bodies, and asking NGOs to submit reports and proposals to the latter while allowing them the right to do so at the former.

Second, NGOs with which this book is concerned are not, or do not, represent narrow sectional interests motivated by financial gain (such as national professional bodies, such as medical associations, or transnational corporations). Instead, they are motivated, implicitly or explicitly, by humanitarian ideals such as the need to protect the environment. They speak for the earth.

Third, NGOs are an alternative source of public opinion. All governments, of course, claim to represent the majority of public opinion of their nations. NGOs represent a different type of public opinion, both more select and yet also more extensive. On the one hand, an NGO represents the views of people who are sufficiently interested in a subject to join an NGO as a way of following up that interest. On the other hand, people interested in the same subject, while they may be in a minority in one nation, will be found in more than one nation. Moreover, NGOs are able to draw upon the resources of non-members such as journalists and scholars to get additional facts and opinion. Also, these NGOs are now so extensive and co-ordinated, especially in New York and Geneva, that they can co-operate among themselves to pool their information and resources, so that at the international level it is very rare for an organization to have to operate on its own. In short, the public opinion an NGO represents is usually both better informed on any issue than the public opinion apparently represented by governments, and it usually extends beyond the confines of any nation's boundaries.

Incidentally, there is another function: that of being a conciliator, or honest broker, between governments. This is not examined here because it is rarely, if ever, performed by environmental NGOs at the international level. This role may yet, of course, evolve for some NGOs. After all, one of the world's most famous NGOs has created a unique role for itself: the International Committee of the Red Cross (ICRC), which devotes much of its attention to being an intermediary, such as by handling the exchange of prisoners-of-war and being a clearing house for information on war victims. Environmental NGOs do not yet have an equivalent of the ICRC.

Various non-governmental organizations around the world are now taking an increasing interest in protecting the Antarctic.[5] In all their activities, the emphasis is upon involving the general public in defending their stake in the Antarctic's future.

The Fifth World and Antarctica

NGO involvement in Antarctic affairs has taken various forms. This section will deal with five illustrations. It is impossible to provide a detailed overview because the situation is so fluid. But at least this will provide some of the flavour of the work.

The Antarctic and Southern Ocean Coalition (ASOC), founded in 1978 to monitor governments' activities in the Antarctic, comprises more than 200 environmental organizations in 35 nations. Its representatives have observed governments' deliberations during the past decade and have provided alternative ideas for the future of the Antarctic. In 1988 ASOC was accepted as a member of CCAMLR. In other words, ASOC is now an important player in the CCAMLR debates. It helps to provide impartial data and fresh thinking on the conservation of living resources. It also regularly produces an international newspaper, *ECO*, which reports on what is happening behind the scenes. ASOC is a loose coalition. It is a clearing house for information – it is not able to direct its members' activities.

One of the US components of ASOC is The Antarctica Project formed in 1982. Its work is based on monitoring, by informing and by lobbying in both domestic and international forums. It holds governments legally, politically and morally accountable for their stewardship of the Antarctic region. The Antarctica Project monitors all important political meetings and government negotiations that may affect the Antarctic's environmental integrity. The Project's activities are implemented on a global basis through the mechanism of the ASOC. Under the ASOC umbrella, it publishes critiques, lobbies internationally and presents ecologically sound alternatives for management of the region. From its Washington DC base, it conducts policy research and legal analysis. It presents testimony and policy proposals to government officials. It prepares articles, books, slideshows, videos and posters suitable for educational and advocacy purposes. It responds to requests for information from across the US and around the world.

Greenpeace, an international environmental organization with three million members worldwide, has been at the forefront of the campaign to save Antarctica. The organization has had a permanent scientific base in Antarctica since 1987, and operates a research, supply and action vessel, the MV *Gondwana*. Greenpeace specializes in direct action, and for 20 years has led campaigns to protect marine mammals, conserve the Antarctic, ban nuclear testing in the Pacific, prevent marine pollution, and bring many other environmental issues to the attention of governments and the public.

Greenpeace has campaigned for Antarctica to be managed as a 'World Park' based on the following four principles:

1. In all decision-making on Antarctica, preservation of the continent's wilderness values should be paramount.

2. There should be complete protection of Antarctica's wildlife.

3. Antarctica should remain a zone of international scientific co-operation, with free exchange of data between scientists of all nations.

4. Antarctica should remain a zone of peace, free of nuclear and other weapons.

> Greenpeace does not believe that mineral exploitation could be carried out in Antarctica without breaching at least some and possibly all of these principles, and is thus adamantly opposed to it. In particular, Greenpeace has argued strongly against the elaboration of the Convention governing Antarctic mineral resources recently completed by the Antarctic Treaty states meeting in Wellington, New Zealand. However, Greenpeace would be prepared to accept exploitation of Antarctic fish and krill, provided that it were done at a rate that did not result in a decline of the stocks. The present administration by the Convention for the Conservation of Antarctic Marine Living Resources has so far not met these criteria.[6]

Greenpeace International has been making ship-based voyages to Antarctica since the early 1980s. In late January 1987, Greenpeace began building the first permanent non-government base in Antarctica as part of its campaign to have the continent declared a World Park. The base consists of private rooms for the over-winterers with communal recreation and service areas. In a way which is markedly different from other settlements in the area all installations and associated waste eventually will be removed from the site, leaving no trace of the Greenpeace presence.

The Greenpeace teams study the Antarctic environment, particularly marine life and the effect of human activity on it. The base has direct satellite communications equipment including a photo-transmitter.

Establishing this NGO presence in Antarctica has been very important for Greenpeace:

> The Antarctic Treaty countries, a very powerful bunch, wanted the world to sit back and let them protect the continent. They said that they had the experience and were doing a good job. We proved them wrong. The first job was to establish a presence by building and occupying the Greenpeace World Park Base. From there, we could document the impact of all the research bases. People around the world could not believe the mountains of trash, diesel fuel and

human waste that poured into the ocean around the bases – and these wastes were produced at scientific stations! We documented oil spills and other industrial-type accidents that were never revealed before. We were able to document abuses and say, 'If this is what a scientific operation is doing, imagine what a large-scale commercial operation would do.' Our documentation changed the way people thought about Antarctica. What seemed like the end of the earth suddenly looked like parts of New Jersey. The key was being there. Antarctica is a frontier setting – no laws, no nothing. And your opinion about it doesn't count for much with some treaty nations if you aren't there. Greenpeace, a private organization, has logged more time on the ice than several major countries. This makes it very difficult for anyone to dismiss our ideas. When a base insisted that certain waste disposal techniques didn't work in Antarctica, we could say, 'But it worked for us. Why won't it work for you?'[7]

Greenpeace International has also protested against the construction of a French airstrip at the French Durmont d'Urville station, claiming that the building and use of the airstrip will disrupt the local wildlife.

Also involved in ASOC is the World Wide Fund for Nature, founded in 1961. It is the largest private international nature conservation organization, with more than three million regular supporters and with national and associate organizations in 27 countries. An example of its work on Antarctica is its lobbying and public educational campaign to win support for its policy on preserving Antarctica's environment. The Australian branch of WWF has the following policy:

1. The strict enforcement of existing protective measures for the Antarctic provided for under the Antarctic Treaty System (ATS).

2. The adherence to strict environmental controls by all those undertaking Antarctic activities.

3. Ensure implementation of international agreements on management of stocks of krill, squid and fin fish which will prevent depletion of these resources.

4. The understanding that the environmental risks of mineral extraction are too great to ever justify such activities.

5. The creation of a new Antarctic conservation regime which incorporates the above points.

6. The opening up of the ATS to greater public participation and scrutiny by allowing observer status to interested countries and international NGOs.

7. Assessment of the Australian Antarctic Territory for the Register of the National Estate.

8. The development of a conservation strategy for the Australian Antarctic Territory.[8]

Finally, there are now several NGOs outside of ASOC which are now taking up Antarctica.

For example, the Amsterdam-based World Association for World Federation, at its July 1988 Council meeting at Turin, Italy, called on the UN General Assembly:

1. to protect and preserve the very constructive achievements of the existing Antarctic Treaty;

2. to democratize decision-making for the Antarctic region so that all nations participating in the General Assembly can participate in it, whether on a basis of state equality or in some other appropriate way, as in the International Seabed Authority;

3. to regard the resources of the Antarctic region as the 'Common Heritage of Mankind' to be used especially to aid low income developing nations but also to help protect the Antarctic environment;

4. to insist that the mineral resources of Antarctica not be exploited unless and until this can be done without significant damage to the fragile Antarctic environment.[9]

The window of opportunity

This is an impressive array of NGO work. But NGOs have no time to lose. The current environmental movement is riding on a wave – but the wave will not last forever. They have to make the most of what, I believe, can be called the Second Environment Wave.

Protecting the environment is nothing new. The first attempt to clean the air over London was made 600 years ago. Therefore, care has to be taken in trying to identify particular beginnings and ends of movements. But I think that we can identify the First Environment Wave of the post-1945 environment movement as beginning in the late 1960s. For by that time, the impact of the post-World War II economic boom was being felt around the world. There was growing concern about the spread of pesticides (DDT), oil spills from oil tankers (such as *Torrey Canyon*), severe river pollution, particularly of the Rhine, Volga, and Danube, the pollution of inland waterways (the most famous case at that time being the US river which had so many pollutants that, in fact, it was a fire hazard!).

Related to the decline in the environmental quality of life, there was concern about whether the post-1945 economic boom could be maintained given the unprecedented rate of consumption of non-renewable resources. The Club of Rome funded the study *The Limits to Growth*. This computer-based study looked at five future trends: increase in food production, availability of non-renewable resources, ability of the environment to absorb pollution, population growth, and the rate of industrialization. Each trend was examined in the context of the carrying capacity of the earth's environment. How many people will it sustain and at what level of affluence? The authors concluded that, if present trends continued, the combination of rapid population growth and increased industrialization would put intolerable strain on the environment, food supply and available energy resources.

The high point of the First Environment Wave was the June 1972 United Nations Conference on the Human Environment (at Stockholm). This was the first of the mega-conferences that came to dominate the political scene in the 1970s. Previous governmental conferences were held in sedate surroundings, largely isolated from the general public and with journalists being fed only bland press releases. The Stockholm Conference changed all that. Around 5,000 people arrived in Stockholm. Governmental delegates were swamped by non-governmental organizations (notably the environment movement) and the Conference was a major media event – made all the more newsworthy by the presence of such notable individuals as Margaret Mead.

In retrospect, the Conference was as successful at consciousness-raising as its organizers had intended. It led to the creation of the United Nations Environment Programme. Governments, in order to be represented at Stockholm, had to have Ministers for the Environment and, therefore, around that time there was a sudden interest in establishing Ministries for the Environment. (I was at that time associated with the attempt to try to create a Ministry for the Environment in the UK; the Labour Government said that it was unnecessary since environmental matters could be taken care of by existing government departments – it was the incoming 1970 Conservative government that established the Department for the Environment on its first day in office!)

But then the First Environment Wave started to collapse. First, the general public is always eager for reassurance that something is being done. The general public is always more concerned about urgent issues rather than important ones. Today's urgent issues take priority over the important ones which may, in the long-run, be far more significant but, in the short-run, have far less direct impact. Thus members of the general public are naturally more concerned about the education of their children, paying mortgages and getting their cars repaired – rather than the state of the global environment, the fact they can be killed several times over in a nuclear war or that the banking system could collapse. Consequently the UN Conference and the establishment of Ministries or Departments for the Environment all contributed to reassuring the general public that something was now being done and that all was well.

Second, in October 1973 there was the Middle East War and the Arab use of oil as a political and economic weapon. Oil suddenly increased in price.

This was of course, at first sight, an advantage to the environment since it encouraged people to be far more careful about the use of petrol. But I think environmental considerations ultimately lost out. People suddenly became worried that the phenomenal material standard of living which was born in 1945 had somehow perished in 1973 in the queue at a petrol station. In 1974, the UN General Assembly met in a very rare Special Session to discuss the creation of a new international economic order. Quite suddenly people were concerned about economic growth rather than protecting the environment. That desire for economic growth continued throughout the 1970s and the 1980s.

Linked to this desire for continued economic growth has been the power of memory. People in Western nations long for a return to the golden age of the 1950s – when they were either children or young adults. The world seemed much more peaceful, pleasant and well-organized. Environmental considerations, of course, were non-existent. Attention was then focused by governments on economic growth. I find as I speak at public meetings nowadays, there is still a longing to return to the stable days of the 1950s – this nostalgia for the past will increase as the years roll by and the past becomes even more attractive. This does not necessarily mean a return to the fashions of the 1950s – simply an attempt to revive the security that people felt at that time.

I believe that mass movements operate in cycles of about seven years. There is nothing mystical about this figure seven. Indeed, the figure seven need not be that specific. It could be six years or it could be nine years. But the cycles certainly do not last as long as say, 15 or 20 years.

The First Environment Wave got underway in the late 1960s and collapsed in 1973.

The first post-1945 peace movement began in 1958. In UK terms, the date is very specific: 16 January 1958, the date of the founding of the Campaign for Nuclear Disarmament (CND). It is amazing just how quickly the movement developed. That January a meeting discussed holding a public meeting at Central Hall, Westminster, on 17 February. Over 5,000 people attended that meeting – a similar meeting held in 1955 had been a flop. Quite suddenly the general public had recognized the dangers of nuclear weapon testing in the atmosphere and so were pressing for nuclear weapon testing to be stopped in particular, and the abolition of all nuclear weapons in general. The nuclear testing issue was clear: above-ground tests put radioactive debris into the atmosphere, which fell to the grass with rain, the grass was eaten by cows, and humans drink milk laced with radioactivity!

The movement was brilliantly successful in making people aware of the dangers of nuclear weapon testing in the atmosphere. In October 1963, responding to public pressure, the US, Soviet and UK governments agreed on the Partial Test Ban Treaty. This stopped nuclear weapons testing in the atmosphere by those three countries (France and China – much smaller players in this matter – were slower to accept the need to test weapons underground).

The peace movement collapsed. People had been reassured that it was now safe to go back to drinking milk. Additionally, the Partial Test Ban Treaty seemed to lead the way to a comprehensive test ban treaty that would stop all nuclear weapon testing. Ironically, there is still no comprehensive test ban treaty and the nuclear weapon nations are still continuing to test weapons.

The next wave of the peace movement began in December 1979. Once again we can be very specific. In that month, NATO agreed to the deployment of intermediate nuclear forces which some people argued would lead to the ability to fight a 'limited' nuclear war in Europe. Governor Ronald Reagan, then campaigning for the White House, talked about the way in which a 'limited' nuclear war could be 'won'. Additionally, the Soviet Union in that same month invaded Afghanistan. This was the first time the Soviet Union had invaded territory outside the unofficial Soviet sphere of influence which arose in 1945. Once again, the growth of the peace movement was remarkable; CND, naturally, was suddenly back in business. Indeed it grew so rapidly that it had to move offices three times to be able to house all the new staff being employed.

The second wave of the peace movement started to taper off in 1987. The US and Soviet leaders agreed to scrap the intermediate nuclear forces which could be used for a so-called limited nuclear war. This was the first time that a whole category of nuclear weapons had been scrapped since 1945. But only four per cent of the world's supply of nuclear weapons has actually been scrapped. Another 96 per cent remains in existence.

Unfortunately the peace movement itself is not around to benefit from that change. The general public remains anxious – but the anxiety is focused on new issues: the environment, drugs and terrorism.

I am inclined to date the Second Environment Wave as from the northern summer of 1988. That hot US summer coincided with media speculation about the greenhouse effect and the hole in the ozone layer. As the UK newspaper *The Guardian* reported in July 1988:

> As the United States buckles under the impact of the worst drought in 50 years, American scientists have confirmed that man-made pollutants are finally beginning to produce the long-predicted and potentially disastrous 'greenhouse effect' on the earth's climate.
>
> Dr James Hanse, the director of Nasa's Institute for Space Studies and an expert on climate, told a Senate committee on the drought: 'It's time to stop waffling so much and say the evidence is pretty strong that the greenhouse effect is here.'
>
> The week's Senate hearings have been crowded with anxious politicians. As the drought has deepened, state governors have warned that food production is being seriously threatened, and farmers are facing penury.
>
> The senators have heard warnings that, unless there are attempts to control atmospheric pollution from industry and agriculture, things will get worse. The warning was repeated last week by Dr Warrick [of the Climatic Research Unit, Colorado].
>
> 'The whole greenhouse effect problem as we go into the next century is going to be the major environmental issue to be faced in the 21st century, and it's just because of the enormity of the problem. The cause gets back to the very essence of the economy, energy use, agriculture, and the whole world's way of life.
>
> 'And that implies that either we change or go ahead with the global experiment in which we are changing the whole environment of the earth.'
>
> Louisiana's Senator Bennett Johnson put it more succinctly. 'We only have one planet. If we screw it up we have no place to go.'[10]

If the seven year cycle is correct, then the second wave will taper off in 1995. For example, the environmental movement will have the same or less – but not more – members in that month than in (say) 1990.

People will still be anxious in mid-1995 but by that time, governments will have taken some partial actions and reassured the general public that all is well.

As to what crises will develop to replace the environment one, I suggest three possibilities. First, the international debt crisis is still not solved; Third World nations will still be having difficulties in 1995 in repaying their debts. Additionally, the pendulum may have started to swing back in favour of economic development. Consequently there will be talk about a new international economic order once again. Second, the population crisis may be at the top of the agenda. This is already beginning to emerge in Australia via the debate over immigration and whether or not Australia should continue to allow foreigners to reside in Australia. Third, the possibility of increased famine may have provoked interest in the food question.

All three potential items are in one way or another related to the environment crisis. That is to be expected. The environment crisis grew partly out of the peace issue, such as the speculation over a nuclear winter following World War III. The tendency in public concern is to slide from one issue into another – rather than a dramatic transfer of concern to a completely unrelated matter. Additionally, all three items are related within themselves. But unfortunately the general public is not able to cope with more than one major crisis at a time.

Some of the NGOs will continue to exist irrespective of the extent of public concern. The Women's International League for Peace and Freedom (WILPF) – which is 75-years-old – has gone through two World Wars, the Vietnam War and the two post-1945 waves of the peace movement. Its membership expands and contracts. But a small core of dedicated workers maintains WILPF's existence; it will still be in operation for the third Peace Wave. Other examples exist in the environment and human rights movements and so on.

Meanwhile, in the interest of saving Antarctica's environment, we need to make the most of the Second Environment Wave. We have no firm indication of just how long there will be a window of opportunity – it could be closed by 1995.

Notes

1. Lyman Cromwell White, *International Non-Governmental Organizations*, Greenwood Press, New York, 1968, p. 3.
2. Jeff Faux, 'The Democrats and the Post-Reagan Economy' in Archibald L. Gillies et al, *Post-Reagan America* New York, World Policy Institute, 1987, pp. 35-6.
3. Michael O'Neill, *The Third America: The Emergence of the Non Profit Sector in the United States*, San Francisco, Jossey-Bass, 1990.
4. A. Pearce Higgins, *Studies in International Law and Relations*, Cambridge University Press, 1928, p. 3.
5. The best overview of environmental NGOs is: Thaddeus C. Trzyna and Ilze M. Gotelli (editors) *World Directory of Environmental Organizations: A Handbook of National and International Organizations and Programs – Governmental and Non-Governmental – Concerned with Protecting the Earth's Resources*, Sacramento, California Institute of Public Affairs, 1990.

6. 'Communications', *The Siren*, UNEP (Nairobi) July 1988, p. 20.
7. Paul Bogart and Sue Sabella, 'What Works: An Oral History of Five Greenpeace Campaigns: Antarctica' *Greenpeace* (Sydney) January 1990, p. 11.
8. *Greenprint for the Environment: A Conservation Agenda for the New Federal Government*, World Wide Fund for Nature (Sydney) 1990, pp. 4-5.
9. Text available from International Secretariat, Leliegracht 21, 1016 GR, Amsterdam, The Netherlands.
10. Michael White, 'Pollution Threatens to Scorch the Earth', *The Guardian* (London) 3 July 1988.

9. No Longer the Neglected Continent

Here to stay

Antarctica is a permanent fixture on the international agenda. The speculation about a 'hole' in the ozone layer and the possible greenhouse effect will in themselves see to that. Antarctica will never return to its earlier status as the neglected continent.

The previous chapters have identified the political and environmental changes gradually having an impact on Antarctica. The changes do not constitute a clear pattern. Indeed, in so far as there has been a debate over Antarctica's legal and political status, the debate has been somewhat confused.

Antarctica's legal and political development has been haphazard. No country has regarded Antarctica as sufficiently important to try to determine consistently Antarctica's legal and political development. Australia, the continent's major claimant, for example, has considered and then dropped the idea of trying to claim all of it; its scientific activities are not proportional to its claim; the USSR and USA – two non-claimants – are more active scientifically than all the claimants combined. Similarly, recent proposals to change Antarctica's political and legal status have all withered fairly quickly. The stalemated Malaysian initiative at the UN is only the most recent of various unsuccessful proposals. In 1975, the then New Zealand Labour Government proposed that Antarctica become a World Park. But a decade later the next Labour Government carried to final drafting (if eventual non-implementation) the treaty which would have regulated but *allowed* mining there.

The remaining chapters of this book are concerned with exploring options for Antarctica. The rest of this chapter contains a format of six questions to help people think about proposals for Antarctica. Chapter 10 deals with the possibility of leaving the Antarctic Consultative Parties in charge of Antarctica. Chapter 11 deals with the Common Heritage of Humankind. The final chapter contains my ideas for declaring Antarctica to be the Public Heritage of Humankind.

Thinking about proposals for Antarctica's future

All legal and political proposals for Antarctica's future need to be viewed against the following system of questions.

1. *What are the values underpinning the proposal? What is the underlying philosophy of the proposal?* For example, the values underpinning the 1959 Antarctic Treaty include: the preservation of Antarctica as a demilitarized zone, controlled by a set number of nations, to be used only for peaceful purposes (especially scientific research) isolated from the turbulent tides of international politics and that these activities should be conducted as far as possible on an international (rather than national) basis, with no changes made to territorial claims.

2. *What use is to be made of Antarctica?* The Antarctica Treaty obviously envisaged Antarctica as being mainly an area of international scientific co-operation. The Convention on the Conservation of Marine Living Resources is an exploitation regime – with a view to not overexploiting the marine resources. In other words, krill may be harvested – but not beyond the point at which the exploitation threatens its future (some have to be left to reproduce) or threatens the future of other life forms which rely on krill as part of the food chain. (The extent to which this is being achieved is, as already noted, now in doubt – but at least the Treaty's stated intention is clear.)

3. *Who or what should own Antarctica?* Ownership/possession/nationality are all important concepts in international law (and municipal law). The Antarctic Treaty contrived (in Article IV) a formula for side-stepping this problem; this has worked brilliantly, but it may now be in some doubt.

4. *Who or what should control Antarctica*? Ownership and control are not necessarily the same. In the domestic context, for example, an individual may 'own' a house, but by renting it out to another person, the owner surrenders some of their 'control' to the tenant. The Antarctic Treaty evaded the ownership issue by focusing upon the issue of control: the Consultative Parties acquired that power.

5. *Who or what should make the detailed operational decisions on Antarctica's use?* The Antarctic Treaty allows these decisions to be made by whatever government is conducting the scientific research. The policy decisions have to conform to the prevailing philosophy of the Antarctic Treaty. But the way in which a base is, for example, established and what research is conducted (if any) is a national decision – which may be done

via negotiations with other governments. If the mining treaty had proceeded, then some of the detailed operational decisions would eventually have been made by transnational corporations and state-run business corporations.

6. *What mechanism is there for peaceful settlement of disputes?* The whole issue of the peaceful settlement of international disputes is now a growth industry in international law and international politics. This reflects the change in the warfare state, whereby the use of military force to settle international disputes is, to say the least, no longer as successful as it was – assuming that aggression was ever as successful as aggressors hoped. Leaving aside the brief and inconclusivc 1982 Falkland Islands conflict, the world ended the 1980s with no new international conflict having broken out during the entire decade. Unfortunately 1990 began badly with Iraq's invasion of Kuwait. This event demonstrated the need for greater attention to the peaceful settlement of international disputes. The Antarctic Treaty, reflecting the era in which it was written, contains no explicit system for the peaceful settlement of disputes. CCAMLR, by contrast, has one major Article (XXV) and the Annex devoted to it. The abortive mining treaty was a much longer document still – and the paragraphs (and the Annex) all devoted to dispute settlement were proportionately longer.

I believe that all proposals for Antarctica's legal and political future should be viewed through the prism of these six questions. All proposals need to be subjected to the first question; the other questions may not necessarily be applicable for all proposals.

Applying the six questions

The purpose of these six questions is, then, to help people (or NGOs or governments) think through proposals being made about Antarctica. Each proposal needs to provide answers for these questions.

Here are five examples of how these questions can be used.

First, what is to be made of Malaysia's proposal (for Antarctica to be declared as a 'common heritage')? Unfortunately the proposal has not – to the best of my knowledge – been set out in sufficient detail to permit a scrutiny on the basis of these questions. In other words, Malaysia has still to think through its proposal. Here, then, is a case of where confusion has coloured the Antarctic debate.

Does Malaysia seek international control (presumably through the UN) of Antarctica? Or does it wish to have Antarctica's wealth (living, mineral, tourist and so on) distributed among the world's nations? Or does it want all exploitation to stop, with the implication that Antarctica becomes some form of 'World Park'? There are crucial differences between these options.

I believe, given Malaysia's appalling record on logging its rainforests, that it is not necessarily motivated by a desire to preserve Antarctica's pristine environment. Malaysia is not known for its international environmental initiatives. On the other hand, it may well resent not being in on Antarctic decision-making – and so wants a say. Given its non-existent record on Antarctic research, it can hardly expect to become an Antarctic Consultative Party. The alternative way of determining Antarctica's future is, therefore, to have it governed by the UN – in which Malaysia is a member-nation.

A second example arises when a person (or NGO or government) may be concerned about Antarctica being preserved as a World Park, but may not necessarily be too worried about how Antarctica was controlled – let alone owned. For example, a treaty could (in theory) be devised which made Antarctica a World Park, completely free from all exploitation, but which left all of Antarctica's affairs controlled by (say) the Consultative Parties. Alternatively, the Consultative Parties could hand over all their power to (say) Australia or the UN. Who or what owned or controlled Antarctica would not be important – providing it was left as a World Park. In other words, this person is worried about questions 1 and 2 (and not worried about questions 3, 4 and 5).

A third example could be a transnational corporation wishing to drill for oil. It would not necessarily be worried about who or what owned or controlled Antarctica (questions 3 and 4) – providing it was given guaranteed access, on the basis of some fair return on investment for drilling the oil (questions 1, 2, 5 and 6). Money has no nationality. Corporations look only for good areas of investment; they are more worried about the security of their investments than the political complexions of governments.

A fourth example comes from Third World nations which resent the fact that international affairs are still influenced by decisions made before their independence and about which they had no say. The Antarctic Treaty came into effect in 1961. The flood of UN membership occurred in the early 1960s. With 110 members in 1962, the UN General Assembly had doubled its size from 1945's total; it now stands at 159. The new members formed the majority and so began to change the UN General Assembly's annual agenda. Their main interests were in decolonization and economic development. The agenda of the UN shifted and widened to include subjects scarcely dreamed of at its inception in 1945 – most notably development, population, food and environment. Consequently, such nations are principally concerned with questions 3, 4 and 5. They are not necessarily worried about Antarctica's use (questions 1 and 2) – providing they have a say in running it.

The final example comes from a person wanting to protect Antarctica's environment against exploitation (questions 1 and 2). Such a person would not be in favour of UN control if this meant that exploitation would take place. Such a person – a 'Fifth Worlder' – is not so much interested in a greater sharing of Antarctica's wealth as in stopping *all* moves to obtain that wealth. Therefore, for those individuals, organizations and governments principally concerned about protecting Antarctica's environment, Malaysia's initiative is not necessarily one worthy of support. Of course, Malaysia may have the capacity to protect Antarctica's environment better than it can its own rainforests – but some caution is still required. This means that there is no automatic coalition between Malaysia and NGOs.

To conclude, some basic questions have been set out which need to be asked of all proposals being made about Antarctica's legal and political future. These reveal just how complicated will be the devising of any future regime. Before dealing with proposals to enhance the international protection of Antarctica, it is necessary to ask if the existing Antarctic Consultative Parties should be left in control of Antarctica.

10. Can We Build Upon the Antarctic Treaty System?

A durable treaty system?

Treaties survive when they are flexible and continue to meet the changing needs of the parties to them. The 1959 Antarctic Treaty is three decades old – and it has worn well; Chapter 2 set out its strengths. Despite its recent problems, the Treaty still has some life in it. The Consultative Parties cannot claim a perfect record in Antarctic environmental protection but acknowledgement must be made of what they have been able to achieve.

This is worth emphasizing as any alternative arrangement to the existing system would need to offer more benefits than the current system. The Consultative Parties have pioneered a nuclear weapon free-zone, with on-site verification, decades before they agreed to such verification within their own boundaries. The 1959 Treaty System has absorbed new parties with minimal friction. The Consultative Parties have negotiated new treaties to meet new problems. CCAMLR has problems – but no treaty can be enforced against recalcitrant nations. The continuing problems for the International Whaling Commission indicate that the Consultative Parties are not alone in having difficulties in convincing member-nations to live up to their international obligations.

Having said all that, however, the 1989 decisions by Australia and France over the mining treaty have created confusion among the Consultative Parties. Australia and France are seeking to protect Antarctica's environment while at the same time not wanting to provide an opportunity to reopen the controversy over territorial claims.[1]

In terms, therefore, of the questions posed in the previous chapter, Australia and France are focusing on numbers 1 and 2, and hoping that the international community (such as Malaysia and its allies at the UN, and NGOs) will not be able to focus on questions 3 and 4.

This chapter follows through the assumption (about which I am myself not too optimistic) that these twin aims can be achieved. How, then, could Antarctica's future protection be determined? Assuming that the ownership and control questions do not need to be addressed, how would Antarctica's fragile environment be safeguarded?

In other words, the Antarctic Consultative Meetings would remain the main policy-setting body. The meetings have adopted over 130 Recommendations covering such matters as the facilitation of scientific research, the exchange of information, telecommunication, meteorology, preservation of historic sites and transport. Additionally, they have drawn up treaties covering flora and fauna (1965), Antarctic Seals (1972), and Marine Living Resources (CCAMLR, 1980). The Antarctic Treaty System could (in theory) go on responding to new issues (or variations on old ones) and adopt new measures (either Recommendations or new treaties).

This chapter looks at three issues which could become the objects of new treaties: tourism; the harvesting of ice; and an Antarctic Environmental Protection Agency. These are the three most obvious matters requiring urgent attention.

One common theme in all these issues is that they would need to be addressed irrespective of who or what controlled Antarctica. In other words, there is no intrinsic reason why the work has to be done by the Consultative Parties. They are the obvious nations to do this work only because they have been responsible for the previous treaties, such as CCAMLR. Additionally, if (say) the UN were to be given control of Antarctic affairs, it would need to carry out this work.

Second, given the Australian and French vetoes of the mining treaty, would the Consultative Parties be willing to go through all the required work knowing that participating governments could change their minds at the last minute? This doubt would be augmented by the realization that governmental and non-governmental pressures will be increased as campaigns continue for Antarctica's internationalization.

Third, it will be increasingly difficult for the Consultative Parties to negotiate new treaties in their previous clubby atmosphere. As Chapters 6, 7 and 8 have argued, Antarctica is now much more in the international spotlight. This means that some nations (especially near election times) may be tempted to win votes at home by adopting 'green policies' on Antarctica thereby disrupting the flow of the negotiations – or even perhaps refusing to sign treaties which they helped negotiate.

A tourist treaty

Tourism in Antarctica is increasing. About 4,000 people went to Antarctica in 1987 and about 5,100 in 1988.[2] Almost all tourists go by ship; Lindblad Travel began cruises there in 1965. Lars Eric Linblad pioneered the Antarctic travel business. He does not believe that responsible tourism is a danger to Antarctic. 'If you take the size of Antarctica, the same as all Europe, including the Soviet part, and compare it with the number of ships and the number of tourists coming to Antarctica, you will realise how very small the numbers are. Australia is smaller than the Antarctic continent and there you can count the ships in the thousands and the tourists in millions.'[3]

Tourist ships carry up to 140 passengers in great comfort, although at considerable cost: a typical fare (twin-share cabin) is close to $US6,000 (approximately A$7,880) for a 10-day cruise, of which six days are spent among the icebergs, glaciers, whales, seals, penguins and other seabirds of Antarctica. Nearly all Antarctic cruises begin from Punta Arenas (Chile), west of Cape Horn, or Puerto Williams (Chile) and Ushuaia (Argentina) on the Beagle Channel in Tierra del Fuego. This is because the rocky finger of the Antarctica Peninsula projects from the frozen continent towards South America, and the gap can be crossed in 48 hours. There are cruises to Antarctica from Hobart, Australia, and Christchurch, New Zealand, but they entail up to 10 days' sailing across seas which are prone to storms – force 12 gales and 15 metre waves.

Robert Raymond, an Antarctic visitor, has summed up the advantages and disadvantages of tourism:

> Today, the growing chorus fears that tourism will ruin the last great wilderness left on Earth. But far from increasing pressure on this delicate environment, where a footprint in lichen may take 100 years to grow back, I believe that tourism may actually help to make the government agencies operating in Antarctica clean-up their acts.
>
> Antarctic tourists are not sheep-like sightseers, but people with strong environmental beliefs – and good connections. Already complaints to their Congressmen by visiting Americans have forced US authorities to allocate tens of millions of dollars to sanitise their huge base at McMurdo Sound. Ministerial orders have been given for the Australian bases to clear up their mess.
>
> Antarctica is not and never will be a place where casual weekenders arrive in their own transport, wander about at will and leave behind what they may. Provided, therefore, that the present style of visitation does not get entirely out of hand, Antarctica may be one of the few places where tourism, in the unaccustomed role of watchdog, may help to preserve the unspoiled character of a wilderness.[4]

Tourism can, therefore, aid in protecting Antarctica. It builds up a constituency of enthusiastic visitors who, having been inspired by their trip, want Antarctica's environment to be protected. Travel books and articles on Antarctica will also help keep the international spotlight on Antarctica.

A warning to tourists

A perpetual concern about tourism is not just what tourists can do to Antarctica – but what Antarctica can do to tourists.

Nowadays all tourists go by sea. But this has not always been so. A quicker method would be to fly direct to one of the bases. Air New Zealand explored this possibility in 1960. The NZ Civil Aviation Division ruled out this possibility because of a lack of passenger accommodation facilities at McMurdo Base, the cost of maintaining fuel reserves at the base to refuel passenger aircraft, and the absence of sufficient emergency facilities to cope with (say) the crash of a large passenger aircraft.

In December 1976, Air New Zealand made a further proposal: one-day flights from Auckland and Christchurch to go over the continent at not less than 16,000 feet. Qantas, the Australian national airline, was also interested in tourism. The first flight took place in February 1977. Air New Zealand organized a total of 14 flights; they were described, at the time, as being 'exceptionally popular'.

All Air New Zealand flights stopped on 28 November 1979. Air New Zealand Flight TE 901, with 257 people aboard, flew into the lower slopes of Mount Erebus. This was one of the greatest peace-time disasters in New Zealand history, virtually every citizen knew at least one person on board.

Recovering the bodies was a nightmare. The aircraft shattered upon impact. Officials had to scoop up the bodies in a hostile environment. Skua gulls, scavengers by nature, pecked at the flesh; they even managed to get at the flesh when placed in polythene bags. One hundred and fourteen substantially intact bodies were recovered and there were another 133 bags containing human remains (in one case, three bags contained parts of one and the same person). Of the total 257 victims, 213 bodies were recovered and identified; the remaining 44 were either irrecoverable or unidentifiable. This was possibly the most difficult retrieval operation carried out in the aftermath of a major air accident – and yet the authorities achieved one of the highest body identification rates (83.4 per cent). Some bodies, or parts thereof, may still be on the crash site. The aircraft wreckage was left on the site. The largest single piece was left intact (about one-third of the entire jet) and it may have some human remains underneath. Police ruled out the possibility of using explosives to break up that piece: an explosion may have revealed more bodies, but it may have just scattered the fragments still more.

The crash was bizarre. Probably no other major disaster in history has been so well photographed at the precise time of the tragedy. Virtually everyone on board had cameras and about 900 frames were eventually processed and found to carry an invaluable record of the flight's progress. The photographs taken while the aircraft was heading straight for Mount Erebus showed everyone smiling and enjoying themselves. The most macabre photograph was taken as the photographer passed from life to death and his or her finger pressed the camera button – the photograph contains a shower of vapourizing fuel droplets spattered on a window.[5]

In other ways, too, Flight 901's last minutes were among the best documented air tragedies. The flight data recorder and cockpit voice recorder were both recovered. Along with the stunning photographs, these data revealed a complete absence of anyone warning that the aircraft was heading straight for a mountain on a clear day.

How could an experienced air crew, employed by an airline with one of the world's best safety records, fly straight into an active volcano on a clear day? Today's passenger aircraft are designed to fly at high speeds and high altitudes along carefully defined flight paths. When Qantas and Air New Zealand went to Antarctica, they flew jumbo jets the way the barnstorming pioneers of aviation used to fly: low and slow and with visual reference to the outside world. Flight 901 was briefed to fly towards the US base at McMurdo Sound, down the broad corridor of sea ice between Ross Island and Mount Erebus and the mainland. But a human error in programming the navigational co-ordinates meant that the flight path of the plane would take it directly over the summit of Antarctica's only active volcano. When the crew descended into the visually deceptive environment of a 'white-out', which rendered Mount Erebus invisible, they saw what their flight plan told them to expect, a white expanse of sea ice, not realizing they had been shifted by the faulty co-ordinates to a collision course with the volcano.

The Flight 901 tragedy echoed through New Zealand's small community for years. Air New Zealand is a major foreign exchange earner; it is one of the world's few airlines to carry each year more passengers than there are people in the country that owns it. It is a New Zealand institution. The initial reaction of the government and airline was to blame the crew for the disaster. There was a demand for a Royal Commission; Peter Mahon, a judge of the Supreme Court of New Zealand, was appointed. He found that the company itself was to blame because of the computer programming error. Because of the findings of the Royal Commission, Justice Mahon was insulted by the then Prime Minister, Sir Robert Muldoon, suffered the slights of an aggrieved New Zealand establishment, and seemed to be hounded into resigning from the Bench.[6]

The crash resulted in various forms of litigation. The most recent case was in July 1988 when US District Judge Harold Greene blamed the airline and crew for the crash. In doing so, he dismissed US$32 million negligence claims against the US Government by relatives of 16 of the 20 crew members. The plaintiffs had claimed that the US Navy air traffic controllers were at fault because they failed to warn the aircraft that it was heading towards Mount Erebus.[7]

Ironically, if the Flight 901 passengers had known fully what we now know, they may not have taken the flight. Only six weeks before the flight, the Antarctic Treaty Consultative Parties, meeting in Washington DC, expressed reservations about the tourist flights because they were operating in a particularly hazardous environment. Antarctic facilities were inadequate for emergency facilities, such as accommodation for (say) 200 stranded passengers, and there were insufficient search facilities to look for missing aircraft. Les Bloxham, a senior journalist with the Christchurch *Press* had written articles warning about the safety implications of these flights.[8] Additionally, Antarctica was a hostile environment – assuming that an aircraft could be landed safely on the ice – in which to wait for a few hours before the passengers could be evacuated from a crash site to a base. The US, which operated the key base in the overflight area, was clear in its opposition to these flights. Additionally, weather forecasting was a key factor in these flights. Antarctica has 30 weather-reporting stations to cover a continent about the size of Europe west of the Urals. Estimating future weather at McMurdo is like getting reports from Moscow, Rome, Prague and Warsaw and trying to forecast London's weather. As the tragedy demonstrated, Air New Zealand had no contingency plan in place to cope with such a disaster.

Despite the Air New Zealand disaster, tourism remains a popular (if small) Antarctic industry. Air New Zealand and Qantas themselves have shown little appetite for restarting their overflights.

A Sydney firm of architects, Helmut Rhode and Partners, has been working for many years on a project which would go beyond ship-based day-trippers. 'Project Oasis, Australia' would be hotel built and operated with the latest technology. Tourists would land on a runway (either in the ice-free area or made from ice) near Australia's Davis Base. Human waste would be burnt – with the energy itself being a source of heat for the hotel. Foreign tourists visiting Australia now constitute a major industry; a quick trip to Antarctica would be an optional extra for them. The overall philosophy would be that of educating tourists about Antarctica – and helping to create a political constituency in favour of safeguarding Antarctica's environment. By making greater use of Australia's Antarctic Territory, this project would, indirectly, also augment Australia's claim to the Territory.

Ambitious proposals such as this one have received little support from environmental non-governmental organizations. Any substantial expansion of tourism carries the risk of increased pollution – even if human waste is burnt or returned to Australia, there is still the risk of accidental pollution. As the continuing human toll of government expeditions reveals, even properly equipped visitors have problems with vehicle accidents, poor weather predictions and the usual working of 'Murphy's Law' so that whatever can go wrong will occasionally go wrong. The Antarctic environment may be fragile – but it can also be pretty savage at times. Rescuing an individual or a small scientific team trained already in survival techniques obviously constitutes an easier task than a Boeing 747 aircraft full of inexperienced tourists.

But, despite the risks – or even because of them – tourists continue to visit Antarctica. The Consultative Parties will need to address this problem. An example of the complexity of the problems occurred in the incident involving the Argentinian ship *Bahai Paraiso* which resulted in the spilling of diesel fuel when it ran aground – it was also carrying about 100 tourists. One requirement should obviously be that supply ships should restrict themselves to supply functions, and not also convey tourists.

Creating a treaty on tourism, then, will not be easy. It could reopen old wounds over territorial claims as claimant nations would try to use tourism as a way of reasserting their claims. Additionally, tourists and tour companies may originate from nations not parties to the Antarctic Treaty. They could apply to join – but they could not be forced to do so.

Harvesting the ice

Could Antarctic ice become a growth industry. Entrepreneurs in Alaska are already pioneering a new growth industry:

> Private businesses are collecting chunks of glacier ice to sell as curios to foreigners, especially the Japanese, and as the base for products like vodka and shampoo. There are also those hoping to market it to people concerned about the purity of their drinking water.
>
> Despite protests by environmentalists, the Alaska Department of Natural Resources has granted 11 glacier ice-harvesting permits allowing the collection of up to 5,000 tonnes of ice a week in the waters of Prince William Sound and the Gulf of Alaska.
>
> However, the US National Park Service and environmental groups, concerned that the infant industry will grow uncontrollably, complain that the entrepreneurs are carving away the Alaskan wilderness for the sake of profit.
>
> 'To me, there's not an aspect of it that isn't wasteful,' said Mr Peter Fitzmaurice, chief ranger for Kenai Fjords National Park, one of the ideal sites for glacier ice harvesting.

> 'You can make ice cubes in your freezer,' he said. 'It's, to me, symbolic of what I see as a bad direction in our development.'[9]

The world is running out of fresh water. Fresh water is a limited resource: only a limited amount of it circulates throughout the planet's ecosystems. Ancient Britons had access to roughly the same amount of water as do the modern Britons. The critical differences are population and usage.

The Antarctic ice cap contains about 90 per cent of the world's ice and 68 per cent of its fresh water. Most of the ice is so thick it never melts (the so-called 'greenhouse effect' may eventually change that). Proposals are circulating for harvesting icebergs that break off from the ice sheet each summer.

In other words, Antarctic ice would not be used as tourist gimmicks – as with Alaska's – but as one way of trying to help ease the world's growing shortage of fresh water.

Gilbert Wallace, an Australian conservationist who has visited Antarctica, has examined the possibilities of harvesting icebergs.

> Even a moderately-sized iceberg, positioned near one of our capital cities, could supply large quantities of fresh water and of power (because of the temperature differential), not to mention the attraction for tourism. A berg two by one kilometre in size, for example, could provide Adelaide with power for one year and water for two. A considerable amount of work has been done to investigate the feasibility of towing icebergs, and the economics are attractive. The actual towing would be reasonably simple, and not too much of the iceberg would melt on the journey. There are other problems however which, while not insurmountable, are somewhat daunting. A useful sized iceberg will have a draft of about 200 metres. Australia's wide continental shelf would make it virtually impossible to float an object of such proportions close enough to the Australian coast for it to be utilised. The nearest location would be about 35 km off Rottnest Island, near Perth. Harvesting an iceberg that far out to sea would be difficult, to say the least. It would seem then that commercial harvesting of icebergs is still a long way off.[10]

Despite this pessimistic assessment, IMBS research company based at Fremantle, West Australia, believes that the icebergs should be harvested.[11] Perth metropolitan water supplies will be 300 million cubic metres short by the year 2040. The company claims that the icebergs are enough to sustain six billion people every year in a continuous cycle – this is the approximate size of today's total world population. The water from the icebergs is so pure that salt has to be added to it for the commercial trials by West Australian soft drink manufacturers. IMBS says towing icebergs would be cheaper than establishing desalination plants or creating a pipeline from the tropical rainfall area in northern West Australia.

A treaty on harvesting icebergs would encounter the type of problems already identified in this chapter.

An Environmental Protection Agency

The Antarctic Treaty has yet another claim to its credit: it has operated for almost three decades without a fixed secretariat. More recent treaties, both on Antarctica and many other matters, have given far more attention to the vexed problem of implementation. Back in the early 1960s treaties were still often seen as a basis of international co-operation, with national governments left to implement them, without some group of international civil servants being established to monitor governmental compliance. Governments have now become more reconciled to the prospect of having international secretariats. They still complain about secretariat costs, are not always prompt in paying their subscriptions and resent secretariat officials being too enthusiastic in their work. But they have learned to live with them.

The 1959 Antarctic Treaty has (by current standards) two omissions. First, it has no fixed secretariat to disseminate information about the Treaty System; to assist the host nations in the holding of Consultative Meetings; to maintain continuity between the Consultative Meetings; and to answer questions and complaints from outside the Treaty System.

Second, the Treaty has no agency to protect or even monitor the Antarctic environment. James Barnes of the Washington DC-based Antarctica Project has summarized the primary functions of an international Antarctic Environmental Protection Agency:

- to undertake investigations and conduct assessments of proposed activities, including logistical support facilities, bases and minerals development;
- to prepare environmental regulations for all proposed activities;
- to conduct inspections and monitoring operations; and
- to report on compliance with rules and regulations directly to Antarctic Treaty governments and other appropriate international bodies.

This can scarcely be said to be a radical idea, given the size and importance of the region and the experience with similar institutional arrangements in so many countries. Yet the Antarctic Treaty Consultative Parties have reacted rather negatively.[12]

To conclude, the Antarctic Treaty System has three sets of problems: Malaysia and its allies are using the UN to undermine the Antarctic Treaty system; environmental problems, such as the greenhouse effect, are presenting new challenges to Antarctica; and merely to maintain their hold over Antarctica, the Consultative Parties will need to get busy creating new treaties covering such matters as tourism, harvesting icebergs and an environmental agency.

Can we build upon the present Antarctic Treaty System? I think it most unlikely. In the following two chapters I explore an alternative system.

Notes

1. 'A Joint Australia–France Proposal in the Form of a Paper Including a Recommendation to the Fifteenth Antarctic Treaty Consultative Parties Meeting', *The Monthly Record* (Canberra) November 1989, pp. 635-40.
2. 'Antarctica's Natural Beauty Helps Peninsular Grow in Popularity', *USA Today* (Washington DC) 29 November 1988.
3. 'Meet Mr Antarctic: Lars Linblad', *The West Australian*, 2 June 1990.
4. Robert Raymond, 'Frozen Assets', *Panorama* (Sydney) April 1990, pp. 33-4.
5. Gordon Vette, *Impact Erebus*, Auckland, Hodder and Stoughton, 1983, pp. 73, 165.
6. Peter Mahon, *Verdict on Erebus*, Sydney, Collins, 1985.
7. 'Crew to Blame for Erebus Air Disaster, US Judge Rules', *The Australian*, 9 July 1988.
8. Ken Hickson, *Flight 901 to Erebus*, Christchurch, Whitcoulls, 1980, pp. 93-5.
9. 'Industry Exploits Alaska's Frozen Assets',*The Australian Financial Review*, 16 May 1990.
10. Gilbert Wallace, 'Icebergs Antarctic Style', *Habitat* (Melbourne) December 1989, p. 23.
11. 'Value in Icebergs', *The West Australian*, 8 May 1989.
12. James N. Barnes, 'Environmental Protection and the Future of the Antarctic: New Approaches and Perspectives are Necessary', in Gillian D. Triggs (Editor), *The Antarctic Treaty Regime: Law, Environment and Resources*, Cambridge University Press, 1987, p. 154.

11. The Common Heritage of Humankind

Words are bullets

The preceding chapters have referred to the common heritage of humankind in the contexts of both the law of the sea and a future Antarctic regime. That concept deserves exploring in more detail.

Words are bullets in international politics. Therefore careful attention has to be given to key phrases, such as 'the common heritage of humankind'. I believe that the Malaysian campaign for Antarctica to be declared as part of the common heritage should be altered. The common heritage concept, although of great intrinsic value, has now become too confused to be useful. In other words, the task is to extract the essence of the concept and to refashion it. This chapter gives the background to the common heritage of humankind and the arguments against proceeding with it in the Antarctic context. The next and final chapter sets out an alternative concept: the public heritage of humankind.

The tragedy of the commons

'The tragedy of the commons' was first popularized by Garrett Hardin over two decades ago.[1] As a Professor of Biology at the University of California, Santa Barbara, he was mainly addressing the issue of over-population. But the essence of his idea applies to any shared resource irrespective of population size. The right to use common land (that is, land not allocated to specific families and individuals) was a well developed right in England throughout the Middle Ages. Peasants could graze their cattle and pigs and collect wood from the local common land. As long as the total activities did not exceed the total carrying capacity of the common land, the activities could continue indefinitely.

But (as an economist would put it) each peasant was a rational actor seeking to maximize self-gain. It was therefore in each peasant's immediate interest to use the common land as much as possible. If all the peasants did this, then in due course they would exceed the land's total carrying capacity and so destroy the land. The 'tragedy of the commons' results when ownership of the resource is held by no one; a large number of users have independent access to the resource; voluntary agreement of all users is required in order to control effectively their collective activities; and total demand for the resource exceeds the supply.

In other words, the land may be 'common' but it also requires some regulation to avoid over-use. Additionally, the political agreement to establish the rules should be negotiated prior to the peasants conducting what would be, in effect, an over-capitalization of their agricultural processes. It is easier to gain agreement on mutual regulation before too heavy an investment has been made. But investment decisions are made by individual actors based on their own financial background and expectations – and not in the context of co-ordination with potential competitors.

The alternative to this type of co-operation is competition: the market system. The market system (in theory) rewards the efficient and punishes the inefficient. It sees the world as a 'dog eat dog' place, where as I have noted, you have to do unto others before they get the chance to do it unto you. It is survival of the fittest.

Unfortunately, short-term economic considerations are ultimately self-defeating. Each actor is not only competing against other actors, but is also taking on the environment – in such a battle the 'winner' becomes a 'loser' because the basis of the actor's ecosystem is destroyed.

This requires a new way of looking at the planet: not as a limitless supply of resources for human use but as something which humans today hold in trust for future generations. This new view of the planet has two implications. First, all humans have to agree on some form of mutual regulation for mutual survival. It is no longer a matter of 'survival of the fittest' but being 'fit to survive'. (This issue is explored in the contexts of the sea and Antarctica in the rest of this chapter.)

Second, the burden of the argument on conservation shifts from conservationists having to justify their arguments to the potential exploiters having to justify their proposed exploitation. Thus the exploiters of Antarctica would need to explain how Antarctica's exploitation will be of benefit to the rest of the world.

Common heritage of humankind

The concept of the common heritage of humankind entered the UN's vocabulary through Arvid Pardo's 1967 speech on the need for the sea-bed to be declared as the common heritage of humankind and for it to be exploited for the benefit of all humankind – rather than just the handful of developed nations which could get there first. The concept has been encountered throughout this book in the context of recent proposals for Antarctica being declared as the common heritage of humankind.

This concept is only the most recent stage in a long debate over the sea: should the sea be beyond any national control or should it be controlled by all nations?

Lawyers all at sea

The sea has long attracted a spirited debate over how it should be controlled (if at all). Wars were fought over which nations could control what water; lawyers battled over the general issue of whether or not the sea could, in fact, be put under any form of national control. The focal point of the debate has been 'freedom of the seas':

> As a general principle of international law the freedom of the seas is less than three and one-half centuries old. It protects the surface of the seas, with the exception of territorial waters, from appropriation or exclusive use by any one state. It was logical that the doctrine should emerge in the age of exploration, when Columbus, Cabot, Vasco da Gama, Ralegh, Drake, Magellan and others began to link the continents of America, Asia, Australia and later, inner Africa, with Europe. The classical legal battle was fought between John Selden, who, as an Englishman in the earlier part of the seventeenth century, tried to protect the interests of a country that was not the leading naval power by advocating a closed sea, and the Dutchman, Hugo Grotius, called the father of international law, who postulated an open sea for his then-dominant maritime country. It is interesting to note that when England later gained naval supremacy, she became the leading champion of the freedom of the seas.
>
> Except in times of war, the major maritime powers have always been supporters of the freedom of the seas, since their navies were able to protect their commercial interests, including, in most cases, their overseas colonies, their merchant fleets, and their fishermen. Between them, they were powerful enough to maintain this freedom as a general doctrine of international law. Until a few years ago they were also the principal advocates of narrow territorial waters – generally a three-mile zone of exclusive sovereignty – the distance a cannonball could be fired effectively in the days of sailing vessels.[2]

Thus, the law of the sea was determined by the world's most powerful nations – which coincidentally also had large navies and so wanted maximum freedom of movement. National politics determined interpretations of the law of the sea. The UK's advocacy of 'freedom of the seas' arose only when it acquired a larger navy.

More recently the territorial sea has been transformed; suddenly developed nations acquired technology to drill the continental shelf and so they decided the land outside the water within their (narrow) territorial sea was, in fact, national territory for the purposes of economic exploitation. They had their cake and also ate it: the world's nations should have narrow territorial seas (so as to permit freedom of maritime movement) but control of the land under the waters beyond national control should be under national control so as to permit drilling for oil and gas. It was only a matter of time before they found a formula to justify national control of the sea-bed as well. If this were achieved, then the technologically advanced nations would still have 'freedom of the seas' while simultaneously exploiting the wealth of the sea-bed via oil drilling rigs and new methods of sea-bed mining for minerals such as gold.

Dr Pardo's 1967 speech led eventually to the UN's decision to establish a Third Law of the Sea Conference. The first UN Conference on the Law of the Sea (1958) adopted the present law of the sea. Some issues could not be agreed upon (such as the width of the territorial sea) and so a second Conference was convened in 1960. This, too, failed to resolve these matters. Dr Pardo's common heritage proposal dealt with the sea-bed but it was first necessary for lawyers to decide where the sea-bed began. This meant a re-examination of the width of the territorial sea. Consequently, Dr Pardo's proposal for the sea-bed in 1967 led to the Third UN Law of the Sea Conference (1974–82).

Two contrasting concepts

There are two contrasting concepts relating to making claims on either objects or geographic areas.

First, there is *res nullius* (an object which has no owner) or *terra nullius* (land which has no owner). The object (*res*) or land (*terra*) goes to the first taker.

It was on this basis that European nations claimed much of Africa, Asia and Latin America. The original inhabitants – not being European – did not count. Thus, a territory did not 'exist' until it was 'discovered' by Europeans such as Captain Cook/Australia and Christopher Columbus/the Americas. Australia, then, was *terra nullius* until the British arrived.

Whales, to take a current controversy, are still to some extent regarded as *res nullius*. Japan and Taiwan continue to operate on a first come, first served basis and resist attempts at curtailing their whaling industries. Some governments have responded by using their national legislation to stop their own nationals from participating in whaling (Australians, for example, would commit a criminal offence in helping those whalers) and in banning the importation of whale products. But Japan and Taiwan cannot be coerced directly by governments into stopping whaling. This illustrates a limitation of international law – nations cannot be forced to accept international obligations.

The second concept is *res communis* (object under common ownership) or *terra communis* (land under common ownership). Malta's 1967 UN initiative on the sea-bed was designed to transfer the sea-bed from the status of *terra nullius* to *terra communis*: the sea-bed was to be declared as part of the common heritage of humankind.

Unfortunately the concept of common heritage is still somewhat unclear:

> The concept of a common heritage for mankind is ill-defined and elusive. The term *res communis* to which the common heritage principle is likened, involves the notion of communal ownership and control of property which is, therefore, not subject to individual ownership or with sovereignty or sovereign rights of particular states. While the notion of a common heritage is, at most, a principle rather than a set of rules, and thus does not entail specific and detailed legal requirements and consequences, it does have certain identifiable objectives. First, it rejects the notion of domination of property or of natural resources either by a particular state as sovereign, or by legal persons as 'owners' of property. Thus it is analogous to the notion of *res communis*. Second, it seeks to safeguard a common heritage for future use, and emphasises conservation of resources, in the sense of sustainable use and exploitation of those resources, and environmental protection. Third, the principle aims for an equitable allocation of the resources and benefits, with particular attention to the needs of developing States. Finally, it contemplates a legal regime to formulate precise rules to give effect to these general objectives.[3]

The status of Antarctica

What, then, is the legal status of Antarctica? Has it been open to conquest on a first come, first served basis – as with Africa, the Americas, Asia and Australia in previous centuries? In other words, does Antarctica have the status of *terra nullius*? The nations which got there first established their territorial claims. As noted in Chapter 2, however, the claims are not recognized by all the world's nations – or even all the Consultative Parties. This lack of mutual recognition means that they have uncertain standing in international law.

Dr Triggs regards Antarctica as *terra nullius*:

> In marked contrast, Antarctica, prior to the intrusion of the notion of a common heritage of mankind, has been viewed as *terra nullius* by the international community and treated as such by states asserting territorial claims. Thus, 90 per cent of Antarctica has been the subject of serious territorial claims for approximately 60 years. The prevailing situation between states, at the time when the notion of a common heritage illumined international consciousness, is thus significantly different from the deep seabed and outer space. It is unlikely that an international tribunal would refuse to acknowledge a consolidated title to territory in Antarctica, simply because the international community has expressed a demand that the resources are to be used in the interests of mankind. This is particularly so in light of the persistent assertions and activities of claimant states with strong, though not exclusive, interests in the territory and its resources.[4]

On the other hand, Antarctica is no ordinary piece of land. There is, for example, the extent to which a nation enjoys 'effective occupation' of the land – and no nation does (despite the activities of Chile and Argentina in trying to 'settle' families there). Moreover, in terms of activities generally, the US and USSR are far more significant than all the claimants combined.

Incidentally, 50 years ago, Admiral Richard Byrd did make a territorial claim on behalf of the US during his third Antarctic expedition (1939-41). The US Government refused to endorse the claim – though both it and the USSR have reserved their right to make a territorial claim at some later time.

The complexity of the US's attitude to making a territorial claim has been described as follows (the same arguments would apply to the USSR):

> The United States obviously would not be willing to fully recognize the claims made by other states and by making a claim of its own the United States would in fact have recognized the right of other states to claim territory on the southern continent. By making a claim of its own, a state will, in fact, accept the very principle of claimability; a national claim to one part implies a recognition of the right of other states to claim other parts. That again implies an exclusion of national pretentions in those other parts where other states may be sovereign. Thus a claim for less than the whole continent does have the disadvantage of limiting the area of operations. Furthermore, if new claims were to be made at a late

> stage, the result would be conflict with prior claimants of the same territory.[5]

More to the point, what territory could the US claim? Most of it is already claimed. The US would either have to be satisfied with the Byrd claim or go after territory claimed by others (all of which are US allies). The US policy is a fascinating blend of high principle and pragmatism. The US actually benefits more from not having its own claim than from having one.

Well: does Antarctica have the status of *terra nullius*? Alternatively, could it be given the status of *terra communis*? I do not know. Indeed, there is no consensus among scholars as to its status. Therein lies one of the limitations of trying to apply the concept of the common heritage to Antarctica.

Limitations of the common heritage concept

To conclude, it is necessary to devise a new concept which embodies the environment-protecting essence of the 'common heritage of humankind' – but which does not use that term. We need to focus the debate on a new regime – without having it sidetracked by the old disputes over what is meant by the term 'common heritage of humankind'. One reason, then, to drop the term 'common heritage of humankind' is its lack of precision:

> Thus the conclusion is justified that the principle of a common heritage does not invoke a legal phenomenon that is clearly understood at international law. Rather, it is a 'label for the bundle of provisions in the agreement creating a new type of territorial status'. To employ the phrase a 'common heritage' in the absence of such a treaty, or readily identifiable customary rule, is to invoke a moral ideal or political aspiration which has no independent legal content.[6]

A second limitation of trying to apply the concept of the common heritage to Antarctica is that it reopens the old controversy over territorial claims. Antarctica is not, then, in the same situation as the sea-bed and outer space: they were not the subject of national claims prior to the UN's declaring them to be parts of the common heritage of humankind. I suppose that the Antarctic claims controversy could be helped along by the International Court of Justice being asked to provide an Advisory Opinion on this matter. But an Advisory Opinion is not a judgement, it would be a lengthy process and could still be ignored –

assuming that the Advisory Opinion was clear (something which has evaded the community of legal scholars).

Third, Malta's 1967 initiative was based on forestalling individual nations and transnational corporations from staking claims to the sea-bed. Malta wanted to avoid the type of scramble for wealth which had occurred a century earlier in Africa. Malta recommended that the sea-bed be internationalized, so as to put it beyond national appropriation, and for it to be exploited for the benefit of all humankind, and not just the minority of technologically advanced nations and transnational corporations. If the sea-bed could be administered smoothly, then this would be a good omen for the governance of the rest of the globe.

But, as the UN Law of the Sea Conference got underway, so coastal nations recognized that the Conference provided them with a unique opportunity to extend their national control over the sea's wealth. The territorial sea was set at 12 nautical miles (many nations had previously claimed only three miles) and there was the creation of an Exclusive Economic Zone (EEZ) extending out for a further 188 miles in which they would have control over the zone's maritime wealth. Forty per cent of ocean space has gone under national control. The push for the EEZ concept was totally unexpected and contrary to Malta's original proposal. In other words, national greed eroded a high-minded humanitarian concept.

Fourth, the concept is an exploitative one as applied to the sea-bed – it is too soon to comment on the fate of the moon in this context. Malta envisaged exploitation on behalf of all the world; the final version is more one of facilitating exploitation for national self-interest. But in either event the concept is an exploitative one. Antarctica, in my view, should be protected against all forms of exploitation – irrespective of the presumed beneficiary.

Finally, the concept is under a cloud. The individual who did so much to publicize the concept – Arvid Pardo, one time Maltese Ambassador to the UN – is himself very critical of the new treaty's incorporation (or non-incorporation) of the concept. He recognizes that the new treaty does have many constructive features. But the extension of national jurisdiction to include the EEZ has meant a dramatic loss of potential scope for the new International Seabed Authority:

> Part XI of the convention, and related annexes, is based on three fundamental assumptions: that deep sea-bed mining would consist almost exclusively of the harvesting of manganese nodules; that this could take place on a commercial scale only in the marine area beyond national jurisdiction; and finally that states and private companies intending to engage in sea-bed mining would be obliged

to co-operate with the future International Seabed Authority because the latter had a near monopoly of commercially exploitable manganese nodule deposits. Voices pointing out that these assumptions were dubious were simply ignored because, I assume, doing otherwise would have meant re-negotiating from the beginning one of the most contentious parts of the convention. Unfortunately, it is now generally recognized that the three basic assumptions in Part XI of the convention do not correspond to reality: deep sea-bed mining will compromise at least the harvesting of polymetallic sulphides and cobalt crusts, and probably also mineral-rich muds, in addition to manganese nodules; deep sea-bed mining will take place not only in the international sea-bed area, but also and perhaps predominantly within the expanded legal continental shelf sanctioned by the new convention. Hence, the International Authority will enjoy no monopoly and will be obliged to develop the minerals of the international sea-bed area in difficult competition with a number of coastal states.[7]

Meanwhile, at the other end of the political spectrum, the right-wing US Heritage Foundation (which has successfully campaigned to stop the US signing the new convention) has argued against the treaty in the context of opposing the UN's work on a New International Economic Order (NIEO):

Perhaps the most widely known effort on behalf of NIEO is the Law of the Sea Treaty. After eight years of fruitless negotiations, the US delegation on April 30, 1982, refused to approve the Draft Convention of the Law of the Sea Treaty. A main reason for US opposition to the treaty was the attempt by the developing nations, represented by a coalition commonly referred to as the Group of 77 [the 'Third World'] to use the negotiations as part of their general effort to establish the NIEO. They insisted that the Sea Law Treaty be based on the notion that resources of the earth, particularly the deep sea-bed mineral deposits, were the 'Common Heritage of Mankind,' to be enjoyed by all, even though the very high cost of developing these resources would be borne by the US and a few other industrial states. This would amount to a massive re-distribution of wealth and technology, conflicting with American concepts of private property, free enterprise, and competition.[8]

A similar criticism has come from a US diplomat (writing in his personal capacity):

> Schemes derived from the 'Common Heritage of Mankind' concept, such as the sea-bed mining provisions of the Law of the Sea Treaty, take regulation one fateful step further. They remove economic activity entirely from private hands in order to establish on the international level large, bureaucratic state enterprises. Like the disastrous white elephants in many developing countries, these schemes would wastefully absorb capital and human talents while becoming a net burden on the international economy.[9]

To conclude, it is unfortunate that the common heritage concept has been eroded and surrounded by controversy. But the reality is that any campaign for Antarctica to be declared as part of the common heritage of humankind will immediately be diverted by discussions on the law of the sea treaty. The concept of the public heritage will be no less controversial – but at least the debate will be focused narrowly upon the merits of that concept and will not go off on a tangent by references to the law of the sea.

Notes

1. Garrett Hardin, 'The Tragedy of the Commons', in Robin Clarke (Editor) *Notes for the Future: An Alternative History of the Past Decade*, London, Thames & Hudson, 1975, pp. 64-7.
2. Wolfgang Friedmann, *The Future of the Oceans*, London, Dobson, 1972, pp. 30-31.
3. Gillian D. Triggs 'The Antarctic Treaty Systems: Some Jurisdictional Problems' in Gillian D. Triggs (Editor) *The Antarctic Treaty Regime: Law Environment and Resources*, Cambridge, Cambridge University Press, 1987, p. 99.
4. Ibid., pp. 103-4.
5. Finn Sollie, 'Trends and Prospects for Regimes for Living and Mineral Resources in the Antarctic', in John King Gamble, Jr. (Editor) *Law of the Sea: Neglected Issues*, University of Hawaii, The Law of the Sea Institute, 1979, p. 195.
6. Gillian D. Triggs, op. cit., p. 103.
7. Arvid Pardo, 'A New Order for the Oceans', *South* (London) May 1984, pp. 73-4.
8. Melanie Merkle, 'The Wayward UN: A Digest of Heritage Studies', *The Heritage Foundation Backgrounder*, Washington DC, July 1983, p. 7.
9. Alan L. Keyes, 'Fixing the UN', *National Interest* (New York) Summer 1986, p. 16.

12. The Public Heritage of Humankind

No easy answers

Antarctica's legal and political future is now more clouded than at any time since 1959. The Antarctic Treaty System has problems from within its own membership, it is being assailed from outside its membership, and crises such as the greenhouse effect seem likely to damage Antarctica's environment.

There are no easy answers to end the Antarctic deadlock.

Most proposals that come from outside the Consultative Parties for Antarctica's future regime are often expressed in general terms, such as a ringing call for it to be declared a world park or wilderness area. Few proposals go into great detail.

There are, however, some signs of hope. These arise out of the events recorded earlier in this book in the context of the increased awareness of the need to protect both the environment generally and Antarctica in particular. This awareness has occurred in tandem with renewed interest in international environment co-operation. This new era was illustrated by J. Alan Beesley, a veteran diplomat, currently Special Advisor on Environmental Affairs to the Secretary of State for External Affairs of Canada:

> To cite only one example, how many of us would have foreseen a decade ago, or even five years ago, the linkage of catastrophic debt problems of developing countries with the preservation of the global ecological system? If Bolivia provides an existing example, Brazil is showing signs of recognition of the need to set an example, on problems involving both the 'lungs of the planet' and the biological diversity which could be lost if these 'lungs' are drained of their life-giving oxygen-generating capacity. Surely this is a new kind of multilateralism, which would press back the frontiers in ways never before contemplated, even in the days of heated controversy during the [1972] Stockholm Environmental Conference.[1]

The 1972 UN Stockholm Conference was a high point in the First Environmental Wave. This book is written during the era of an emerging Second Environmental Wave and is seeking to capture some of that wave's vision and enthusiasm. The crisis over Antarctica contains opportunities for action.

The challenge, then, is threefold. First, there is a need to make the most of the current public awareness about the environment in such a way as to draw particular attention to the problem of saving Antarctica. Second, there is a need to avoid alienating the Antarctic Consultative Parties. We cannot function with them and yet we cannot function without them! Little is to be gained, for example, by continuing to call for Antarctica to be governed by the UN; the Consultative Parties are not going to back down and lose face. A new administrative arrangement is required that replaces the Consultative Parties but does not entail Antarctica's transfer to the UN. Third, Antarctica should be preserved as some form of World Park based on the environment-protecting essence of the concept of the common heritage of humankind.

This chapter sets out the proposal for Antarctica to be declared the Public Heritage of Humankind. The concept of 'public heritage' has not previously received the type of application to Antarctica as set out in this book. In essence I have tried to create an alternative to the 'common heritage of humankind' which would be exclusively a conservation regime. A further innovation is the proposed Antarctic Public Heritage Agency. The proposal also contains an application of the Binding Triad form of governance. This idea has been proposed in the context of UN reform; I am using the idea in the somewhat different context of Antarctica.

The concept of the Public Heritage

Humankind will last if it puts the environment first. This means that the environment is not just one of several variables to be considered when making political decisions – *its protection should be the basis against which all the other variables have to be assessed*.

The concept of the Public Heritage of Humankind is derived partly from the common heritage of humankind. But there are some crucial changes. The word 'public' is drawn from the law of the Roman Empire: the public trust. As G.M. Bates has explained, the 'theory is really very simple: that natural resources such as rivers, the seashore, air and wilderness areas are the common property of mankind, to be held by the Crown or government 'in trust' for the common benefit and use of the public generally.'[2] Certain areas are so important that it would be unwise to allow them to become privately owned; as a gift of nature they should be made available freely to all people irrespective of economic or social status; and governments should promise their use for all citizens rather than have them used for only a minority of citizens.

'Public trust' may not be part of the ordinary vocabulary – but tourists are well aware of its benefits. In the US, for example, recognition of such a trust has led to the protection of national parks and wildlife. There has also been the recognition under this doctrine that recreational and aesthetic rights are just as worthy of protection as more traditional economic and property rights. Additionally, the government cannot abandon its legal authority and responsibility to manage and protect public lands.

Two points should be noted in this context. First, the public trust doctrine is still only emerging in domestic law. It carries more weight, for example, in the US than in Australia.

But at least the emergence of the new doctrine of public trust in domestic environmental law means that the proposed concept of Public Heritage is not entirely unprecedented. It flows from some thinking that is already underway in the concept of domestic law. Indeed, as tourism becomes an even bigger industry and the environment movement increases its political muscle, so domestic examples of public trust tracts of land and buildings will increase.

Second, as Bates points out, 'The acceptance of the public trust doctrine does not, of course, mean that no development on public lands can ever be carried out . . . The trust doctrine simply demands a balance between what will be gained and what will be lost.' For example, a forest may be logged or a river dammed.

The concept of the Public Heritage in this chapter calls for the total avoidance of any fishing or mineral exploitation of Antarctica. This is a major departure from the domestic use of 'public trust'. As Dr Bates knows very well (as a Tasmanian 'Green' State parliamentarian) logging in rainforests is very controversial; a total ban is easier to advocate. It removes any promise of exploitation – and so companies strike an area off their list of potential economic activity. There is no protracted debate over the 'sustainable utilization' of a resource. Additionally, a total ban is easier to monitor.

Consequently, this book's proposal is not a direct transfer of 'public trust' from the domestic sphere to the international one. Hence I have coined the phrase Public Heritage rather than proposed the use of 'public trust' in Antarctica's context.

Antarctica represents a unique application of the concept of Public Heritage because of the problem of 'ownership'. First, Stonehenge is British; no one challenges that. The UK's Antarctic claim, by contrast, is not universally accepted. Second, the administrative arrangements for domestic areas of public trust vary from one nation to another. Protecting the Great Barrier Reef, for example, was complicated by

Australia's federal structure and the lack of co-operation from the Queensland government; the UK does not, by contrast, have a federal structure.

Chapter 9 set out six questions as an aid to thinking about proposals for Antarctica's legal regime. Those questions are now applied to the concept of the Public Heritage of Humankind.

1. Values: there are seven values underpinning the concept.

First, Antarctica is not available for any territorial claims: it belongs to all humankind. All of humankind, as far as is practicable, should be involved in Antarctic decision-making.

Second, the debate over Antarctica has to be conducted not only in the context of today's concerns but also on the basis that this is a vital matter for future generations. Antarctica represents a legacy which has to be conserved for future generations.

Third, no economic exploitation at all may take place within the zone of the Antarctic Convergence/south of Latitude 60 degrees South.

Fourth, Antarctica's preservation has to be viewed in the context of the planetary environmental agenda. Antarctica's preservation will in itself help the international effort to protect the environment – and the method used should itself inspire other efforts elsewhere.

Fifth, Antarctica's protection should be done on the basis of stressing two forms of global interdependence. One is the interdependence of nations, recognizing that all nations rely to some extent on other nations for essential ingredients in their economies, living standards, or security. The other form of interdependence is the complex interactions among the issues of population, energy, resources, pollution, food, money supply, economic growth and development. A problem in one sphere worsens problems in others.

Sixth, Antarctica should remain as essentially a zone of wilderness whose main human visitors are scientists and tourists (under supervision).

Seventh, Antarctica should remain demilitarized (as per the 1959 Antarctic Treaty) and no nuclear power stations should be installed there.

2. Use: the primary use of Antarctica would be scientific research conducted on the basis of international co-operation. Ship-based tourism would be permitted, but no fixed hotels. No exploitation of Antarctic marine life or minerals or icebergs would be permitted.

3. Ownership: ownership would be vested in humankind – both current and future generations. This would be a variation of the principle of *terra communis* encountered in Chapter 11. That principle provided for common exploitation of a resource; the Public Heritage of Humankind is not available at all for exploitation.

This will, naturally, require the seven claimant nations to cease making their claims. This is going to be one of the hardest problems in creating any new regime for Antarctica. But, then, this has always been one of Antarctica's problems. Although partly settled – or, more accurately, set aside in 1959 – it will loom large again as the 1959 Treaty encounters the difficulties identified earlier in this book. Without denying those difficulties, this chapter later on provides some inducement to the claimant nations to surrender their claims.

To 'own' Antarctica on behalf of the world's present and future population, a new international organization needs to be established: the Antarctic Public Heritage Agency (APHA).

4. Control: The Antarctic Consultative Parties have painted themselves into a corner in their opposition to any form of UN control over Antarctica. Is it possible to internationalize the control over Antarctica without relying on the UN system of one nation–one vote?

The Binding Triad may be a way out of this problem. The Binding Triad is part of a set of proposals for reforming the UN by the New York-based Center for War/Peace Studies, headed by veteran journalist Richard Hudson. Hudson's work has been focused mainly on UN reform but the essence of this proposal could form the basis of a new system for controlling Antarctica – or running any other international organization for that matter.

The UN General Assembly consists of all UN member-nations (currently 159) and adopts non-binding resolutions. The resolutions often have an air of unreality about them: they are expressions of opinion about what the majority of government would like to see done rather than based specifically on what could be done. General Assembly sessions have become resolution-producing factories. Since the resolutions are non-binding, governments know they will not be obliged to implement the resolutions (such as ceasing trade with South Africa). Additionally, the General Assembly is based on one nation–one vote and so the majority of the Third World can use their voting muscle to push through a resolution – even though they lack the political, economic and military power to ensure that it is carried out.

Hudson's ambitious response is to recommend that General Assembly resolutions be binding and that the powers of the General Assembly be increased. But to give the General Assembly more responsibility means having to make it feel more responsible for its actions. One part of Hudson's reform package is the Binding Triad.[3]

The Binding Triad proposal means that each resolution would need to attain a two-thirds majority in three tiers of voting. One tier is the UN's present system: one nation–one vote. The second would be based on each nation's contribution to the UN budget (a formula devised from each nation's gross national product) and so a successful endorsement at this stage would mean that the world's wealthiest nations supported it. The third tier would be based on each nation's population and so a successful endorsement would mean that the world's most populous nations (which are often among the poorest in GNP terms) supported it. In summary, each resolution would need to be supported by two-thirds of the world's nations, two-thirds of the world's richest nations, and two-thirds of the most populous nations.

Hudson's package of reforms includes ideas on UN Charter amendment and enhancing the General Assembly's powers. These are not relevant in this book's context, since it is proposed here that Antarctica should be administered outside the UN system.

All the world's nations would be invited to join the Antarctic Public Heritage Agency, both UN member-nations and those (notably Switzerland) which are not UN member-nations. Voting would be done on the basis of the Binding Triad.

International non-governmental organization (NGOs) would be eligible to have the consultative status which they enjoy at UN meetings: to attend, to observe, to make statements (with the chairperson's permission), but they would not have the right to vote.

5. **Operational Decisions**: All major operational decisions would be made by APHA. In other words, national governments and NGOs may continue to send their scientists to Antarctica and ship-based tourists may visit there, but all such trips would need to get the permission of APHA.

In due course, it is hoped that APHA would be able to mount its own international scientific expeditions.

6. **Peaceful Settlement of Disputes**: This is now a standard feature of many international agreements (including the two Antarctic treaties on living resources and mineral activities). This point need not, then, be laboured in this context.

Some attention should, however, be given to one project which will help this process: satellite verification. This work has been pioneered by Colonel Howard Kurtz (US Air Force, retired) and his late wife, the Reverend Harriet Kurtz. The essence of the Kurtz agenda is that the major obstacle to disarmament is not being addressed: nations feel insecure with all their arms, but even more insecure at the thought of disarmament. Nations will not disarm until there is some way to provide security through a common global system. Satellites can be used both to verify disarmament treaties and monitor potential conflicts as well as to build an international climate of friendship based on mutual co-operation in economic and social affairs.

Satellite technology (one of the fastest growing areas of space research) provides, ironically, ways of both enhancing and eroding national security. The American Landsat satellite, in one orbit of the earth in approximately 103 minutes, invades the privacy of every sovereign nation in its path 115 miles wide. Once every 18 days this one Landsat satellite inspects in increasingly minute detail the sovereign territory of every other nation on the planet. During the 18 days it takes one Landsat satellite to photograph the entire earth, its instruments are recording what is seen and transmitting the data for permanent storage in ground computers for sovereign US territory only 1.8 per cent of the time; 98.2 per cent of this time Landsat is gathering intelligence or information from territory that is foreign to the United States. US multinational corporations are receiving preferred access to the energies and resources and potential intelligence to gain economic dominion over these nations. But, according to the Kurtzes, Landsat could become the leading instrument of a transcendant new US national commitment to help all nations in their struggles for their own future security and progress. The Kurtzes' proposal to use satellites for monitoring purposes was picked up in 1978 by then-President of France Valery Giscard d'Estaing, who recommended a UN International Satellite Monitoring Agency. Although then opposed by both the United States and the Soviet Union, the idea has gained support.

The USSR, under its new leader, has become far more interested. Meanwhile, the US Government (which continues to be opposed to an international verification agency) is considering its own system. The US National Aeronautics and Space Administration (NASA) has devised a project entitled Mission to Planet Earth. This would consist of a series of satellite flights designed to monitor the earth with instruments that measure such vital signs as temperature, winds and atmospheric chemistry. These readings would add to the knowledge gained from high-resolution photography alone. The object is to understand the planet's dynamics well enough to anticipate ecological disasters – and find ways to forestall them. Detectors would monitor the shrinking of the tropical rainforests as well as of the polar ice caps (a possible consequence of global warming).[4]

Walter Dorn, a scientist at the University of Toronto, has proposed an International Monitoring Agency for the Arctic:

> Through the agency, participating nations could also share the cost and the equipment required for joint Arctic monitoring. Over time, the range of information gathered would expand from environmental and scientific data to include military activities, both above and below the surface. At first, the agency would monitor only international territory, covering the ice and surface waters of the

> polar basin. Subsequently it could expand its surveillance to include territories within the 200-mile exclusive economic zones and the 12-mile coastal zones. At the request of a state, the agency could also carry out specific monitoring tasks within national borders.[5]
>
> An international Arctic monitoring agency would be a catalyst for international co-operation. It would help prevent further militarization of the Arctic by helping to remove the underlying suspicion and mistrust behind militarization. It would increase our knowledge about the region and would save governments money by allowing them to share the costs of Arctic monitoring. The positive effects of an open Arctic monitoring system would be felt far beyond the circumpolar North. At this special juncture in history, while Arctic policy is being cast and East–West relations are favourable, joint Arctic monitoring is an opportunity we must consider seriously.[6]

Dorn's conclusions would also apply to a similar agency run by APHA dealing with Antarctica.

Benefits of the Public Heritage to the Consultative Parties

First, it will provide a general guarantee that no one is secretly preparing to exploit the Antarctic. Thus, for example, it could be argued that the Treaty's Consultative Parties negotiated a mineral regime not because they wanted to keep the Third World out or because they want to monopolize Antarctica's wealth, but rather because they did not trust each other not to take advantage of some small gain. The Public Heritage concept represents a self-denying ordinance. The best way to allay suspicions is simply for all nations to agree to maintain Antarctica as a wilderness area free from economic development.

Second, a blanket prohibition on all economic development will deter transnational corporations from even listing Antarctica as a potential area of activity. It will mean Antarctica will never appear on the list of corporations' long-term projects. As long as there is a hint of a chance of being able to exploit the resources, corporations will lobby for the right to do so – as has been seen in the UN debates over the sea-bed. A blanket prohibition will thus save the Consultative Parties from being lobbied by transnational corporations anxious to get access to Antarctica's resources.

Third, it is necessary for the Consultative Parties not only to act above suspicion but also to appear to be above suspicion. Thus, the Consultative Parties have apparently done little in respect of acting upon France's clear violation of the Agreed Measures for the Protection and

Conservation of Antarctic Fauna and Flora – with its plans to flatten islands near its base, Dumont d'Urville, in order to construct an airstrip. The failure to act upon breaches of an existing legislation does not inspire confidence for responsible environmental management. As long as there are grey areas, such disputes will exist. A blanket prohibition automatically excludes debates over wording and spirit of agreements.

Fourth, by surrendering their claims, the seven claimant nations could cease having to devote effort trying to maintain those claims. Those nations could then focus on the more important task of protecting Antarctica's environment.

Fifth, the Public Heritage proposal is a way of coping with the internal stresses of the Consultative Parties. The Consultative Parties are not, for example, united in their opposition to some form of internationalization of Antarctica.

The Peoples Republic of China (PRC) has a view not so different from that of Malaysia's:

> The secrecy of the minerals regime negotiations and the scarce publicity concerning Chinese Antarctic activities make a reliable examination of the Chinese position difficult. It is clear, however, that China supports the ideas of the 'common heritage of mankind', as it has stressed the vast potentialities of the Antarctic continent to 'serve peace and the common interests of mankind'. Despite the lack of clear interpretation of the common heritage principle, China's statement echoes its previous recognition at the Third United Nations Conference on the Law of the Sea that access to 'global commons', such as outer space and the international seabed, should be governed in the interest of the international community. The PRC has not made any territorial claims but neither has it officially recognized any other Treaty member's claims, though it was reported that New Zealand had proposed a station site for the PRC in return for Chinese recognition of the New Zealand claim. In fact, several Chinese legal scholars have argued that territorial claims on Antarctica based on discovery, occupation, and the sector theory cannot be recognized as valid and that the continent should be jointly owned by mankind. The PRC maintains a fairly internationalist line and favors U.N. involvement.[7]

The Public Heritage proposal provides a compromise for the Consultative Parties. It creates international control (which would please China) but not specifically under the UN (therefore requiring no backdown from such countries as Australia, the UK and US).

Sixth, the proposed Antarctic Public Heritage Agency would not have a difficult – or expensive – set of tasks. Its overall aim would be to preserve Antarctica as a wilderness, to ensure freedom of scientific endeavour, and the regulation of ship-based tourists. Much of this is already being done separately by the Consultative Parties. Since the Agency would be funded by international subscription, then there would not be a dramatic increase in costs – except for the cost of installing the international satellite verification system. A new treaty would be required and this would replace the 1959 Antarctic Treaty. Since all exploitation would be stopped, then CCAMLR would have to be superseded. No further consideration could be given to a mining treaty.

Finally, there is Suter's Law of Politics: the further a government is away from a problem, the more definite its views! For example, the Australian government is far more strident in its views on *apartheid* in South Africa than it is on the treatment of Aborigines in Western Australia – when there is a higher per capita rate of Black deaths in police custody in Australia than in South Africa. Governments in the Consultative Parties under pressure at home to protect the environment may find it easier to protect Antarctica's environment than to introduce radical reforms at home. It would be easier to create and implement the Public Heritage proposal than to stop all rainforest logging in Australia or clean up England's rivers (said to be among the most polluted in Western Europe).

Benefits of the Public Heritage to the planet

The 1959 Antarctic Treaty was a pioneering effort for its era. Any replacement for that Treaty should maintain that Treaty's record for creating precedents.

One benefit of the Public Heritage proposal would be the explicit decision to preserve a substantial part of the globe – perhaps as much as 10 per cent if the Antarctic Convergence is included – as a wilderness protected against exploitation.

Second, the Public Heritage proposal would demonstrate the value of the Binding Triad system of voting.

Third, the proposal would provide a concrete way of demonstrating the end of the Cold War by creating new and inspiring opportunities for US/Soviet co-operation. The proposed Antarctic Public Heritage Agency would be the first international institution established since the end of the Cold War and so would be a litmus test for US/Soviet co-operation. It would be a new beginning in a pattern of international institutions freed from the usual First World/Second World/Third World factions. Additionally, the new Agency would permit the 'ex-enemy States' (as Japan and Germany are called in the UN Charter) to attain senior positions in international institutions freed from the stigma of World War II.

Campaigning for Antarctica – a job for life

The May 1989 decisions by Australia and France not to accept the Antarctic Mining Treaty was a turning point in Antarctica's legal and political evolution. A stalemate now exists between the Antarctic Treaty Consultative Parties with Australia and France advocating the creation of some form of world park but being blocked by most of their partners. Meanwhile there is growing public concern about the threats to Antarctica's environment. The challenge is to convert public concern into public policy.

The proposals set out in this chapter are visionary and ambitious. They are, to varying extents, probably unacceptable to most of the Antarctic Treaty Consultative Parties. For example, the two Parties most sympathetic to preserving the Antarctic as some form of world park – Australia and France – would not accept the element of international ownership and control. (China would presumably not be so hostile to the proposal.)

Environmental non-governmental organizations – the Fifth World – would, however, find these proposals far more acceptable because they could recognize their philosophy enshrined in them. I therefore hope that they could use the proposals as the basis for their campaigning.

Why propose something which is – at present – politically unacceptable? First, the campaign against the mining treaty provides an answer: something which is politically unacceptable today may become politically acceptable tomorrow. If the Fifth World had decided not to oppose the mining treaty – since it was obviously politically acceptable at the time – then the treaty would have been accepted by Australia and France: a self-fulfilling prophecy. The Fifth World decided, however, to campaign against the treaty – and the campaign bore fruit.

Second, it is the Fifth World's responsibility to enlarge the vision of governments. After all, it was the Fifth World which campaigned in the 1960s to arouse public opinion about the dangers of environmental destruction – and this work led to the 1972 UN Conference on the Environment. Governments respond to campaigns. Consequently this book is designed to inject some fresh thinking into the debate over Antarctica's legal and political future.

When it comes to saving Antarctica it is much better to try something grand and fail – rather than to try nothing and succeed.

Third, the past two years have seen remarkable changes due to Mr Gorbachev's *perestroika* and *glasnost*. He represents a 'dam-buster' – a vast political 'lake' had built up behind a 'dam' of stale thinking about centralized economic planning and the stifling of political dissent. Mr Gorbachev has broken that 'dam' and a river is flowing along (well beyond the control of Mr Gorbachev) into parched areas. The 'water' will give new life to previously arid areas of political 'desert', such as international co-operation in protecting the environment. The Fifth World needs to be ready with suggestions.

Disarmament negotiations illustrate this point. For most of the 1980s, the US and Soviet disarmament negotiators met, discussed and agreed on nothing. There was not the political will in the US and Soviet leaders for any disarmament agreement. Quite suddenly, the political will existed simultaneously for an agreement to scrap in Europe the intermediate-range nuclear forces. The negotiators had then to produce the technical details – running into 100 pages! – to convert the political will into a specific treaty.[8] In other words, the negotiations had to do the technical work and have the details drafted in readiness for a change in the political situation.

The Fifth World has two simultaneous tasks: campaigning for changes in the political will over Antarctica while simultaneously having a package of proposals ready to present to politicians once they are willing to move beyond their present positions.

Finally, Michael Parfit, an American writer, spent several months in Antarctica collecting material for a travel book. His book captures some of the excitement of life in Antarctica:

> Here was the first modern continent, described by science, settled without nationality, used without waste, governed without borders. Antarctica was full of squabbles, but they all stopped at one point of agreement: the place was too precious to waste with war. Antarctica allowed us to imagine the world at peace.[9]

The next few years will contain threats to that ideal. There are no easy answers to these threats.

The Public Heritage concept will help supply new thinking. Ideals in politics are rarely achieved – but the pursuit of them determines history. The Fifth World is striving to protect Antarctica. It may have to accept something less than the Public Heritage ideal – such as fishing in the Antarctic Convergence – but its campaign will broaden governmental thinking on international environmental protection, reaffirm the value of wilderness areas, remind people that through the bleak years of US/Soviet hostility there was no Cold War at the South Pole, and demonstrate that it is possible to create a global partnership to protect Antarctica.

Notes

1. J. Alan Beesley, *New Frontiers of Multilateralism: The John W. Holmes Memorial Lecture*, Hanover, New Hampshire, The Academic Council on the United Nations System, 1989, pp. 8-9.
2. G.M. Bates, *Environmental Law in Australia*, Sydney, Butterworths, 1987, pp. 298-300.
3. Richard Hudson's ideas are presented quarterly in the newsletter *Global Report*, New York, Center for War/Peace Studies.
4. 'Taking the Earth's Vital Signs: NASA Seeks to Become a Major Force in the Campaign to Save the Planet', *Time Australia*, Sydney, 4 June 1989, p. 56.
5. Walter A. Dorn, 'An International Monitoring Satellite for the Arctic', *Information North: Newsletter of the Arctic Institute of North America*, University of Calgary, January 1989, p. 1.
6. Ibid, p. 4.
7. Wei-chin Lee 'China and Antarctica: So Far and Yet so Near', *Asian Survey*, University of California, 1990, p. 583.
8. Treaty Between the USA and USSR for the Elimination of Their Intermediate-Range and Shorter-Range Missiles. *Arms Control and Disarmament Agreements: Texts and Histories of the Negotiations*, Washington DC, US Arms Control and Disarmament Agency, 1990, pp. 350-454.
9. Michael Parfit, *South Light: A Journey to Antarctica*, London, Bloomsbury, 1988, pp. 291-2.

Further Reading

Dean Edwin Abrahamson (Editor), *The Challenge of Global Warming*, Washington DC, Island Press/Natural Resources Defence Council, 1989.

F.M. Auburn, *Antarctic Law and Politics*, London, Hurst, 1982.

Creina Bond, Roy Siegfried and Peter Johnson, *Antarctica: No Single Country, No Single Sea*, London, Hamlyn, 1979.

Jonathan Chester, *Going to Extremes: Project Blizzard and Australia's Antarctic Heritage*, Sydney, Doubleday, 1986.

Richard Fifield, *International Research in the Antarctic*, Oxford, Oxford University Press, 1987.

Stanley Johnson, *Antarctica: The Last Great Wilderness*, London, Weidenfeld and Nicolson, 1985.

Geoff Mosley, *Antarctica: Our Last Great Wilderness*, Melbourne, Australian Conservation Foundation, 1986.

Michael Parfit, *South Light: A Journey to Antarctica*, London, Bloomsbury, 1988.

M.J. Peterson *Managing the Frozen South: The Creation and Evolution of the Antarctic Treaty System*, Berkeley, Los Angeles, University of California Press, 1988.

Gillian Triggs (Editor), *The Antarctic Treaty Regime: Law, Environment and Resources*, Cambridge, Cambridge University Press, 1987.

Appendix I: The Antarctic Treaty

The Governments of Argentina, Australia, Belgium, Chile, the French Republic, Japan, New Zealand, Norway, the Union of South Africa, the Union of Soviet Socialist Republics, the United Kingdom of Great Britain and Northern Ireland, and the United States of America,

Recognizing that it is in the interest of all mankind that Antarctica shall continue forever to be used exclusively for peaceful purposes and shall not become the scene or object of international discord;

Acknowledging the substantial contributions to scientific knowledge resulting from international co-operation in scientific investigation in Antarctica;

Convinced that the establishment of a firm foundation for the continuation and development of such co-operation on the basis of freedom of scientific investigation in Antarctica as applied during the International Geophysical Year accords with the interests of science and the progress of all mankind;

Convinced also that a treaty ensuring the use of Antarctica for peaceful purposes only and the continuance of international harmony in Antarctica will further the purposes and principles embodied in the Charter of the United Nations;

Have agreed as follows:

Article I

1. Antarctica shall be used for peaceful purposes only. There shall be prohibited, *inter alia*, any measures of a military nature, such as the establishment of military bases and fortifications, the carrying out of military manoeuvres, as well as the testing of any type of weapons.

2. The present Treaty shall not prevent the use of military personnel or equipment for scientific research or for any other peaceful purpose.

Article II

Freedom of scientific investigation in Antarctica and co-operation toward that end, as applied during the International Geophysical Year, shall continue, subject to the provisions of the present Treaty.

Article III

1. In order to promote international co-operation in scientific investigation in

Antarctica, as provided for in Article II of the present Treaty, the Contracting Parties agree that, to the greatest extent feasible and practicable:

(*a*) information regarding plans for scientific programs in Antarctica shall be exchanged to permit maximum economy and efficiency of operations;

(*b*) scientific personnel shall be exchanged in Antarctica between expeditions and stations;

(*c*) scientific observations and results from Antarctica shall be exchanged and made freely available.

2. In implementing this Article, every encouragement shall be given to the establishment of co-operative working relations with those Specialized Agencies of the United Nations and other international organizations having a scientific or technical interest in Antarctica.

Article IV

1. Nothing contained in the present Treaty shall be interpreted as:

(*a*) a renunciation by any Contracting Party of previously asserted rights of or claims to territorial sovereignty in Antarctica;

(*b*) a renunciation or diminution by any Contracting Party of any basis of claim to territorial sovereignty in Antarctica which it may have whether as a result of its activities or those of its nationals in Antarctica, or otherwise;

(*c*) prejudicing the position of any Contracting Party as regards its recognition or non-recognition of any other State's right of or claim or basis of claim to territorial sovereignty in Antarctica.

2. No acts or activities taking place while the present Treaty is in force shall constitute a basis for asserting, supporting or denying a claim to territorial sovereignty in Antarctica or create any rights of sovereignty in Antarctica. No new claim, or enlargement of an existing claim, to territorial sovereignty in Antarctica shall be asserted while the present Treaty is in force.

Article V

1. Any nuclear explosions in Antarctica and the disposal there of radioactive waste material shall be prohibited.

2. In the event of the conclusion of international agreements concerning the use of nuclear energy, including nuclear explosions and the disposal of radioactive waste material, to which all of the Contracting Parties whose representatives are entitled to participate in the meetings provided for under Article IX are parties, the rules established under such agreements shall apply in Antarctica.

Article VI

The provisions of the present Treaty shall apply to the area south of 60° South Latitude, including all ice shelves, but nothing in the present Treaty shall prejudice or in any way affect the rights, or the exercise of the rights, of any State under international law with regard to the high seas within that area.

Article VII

1. In order to promote the objectives and ensure the observance of the provisions of the present Treaty, each Contracting Party whose representatives are entitled to participate in the meetings referred to in Article IX of the Treaty shall have the right to designate observers to carry out any inspection provided for by the present Article. Observers shall be nationals of the Contracting Parties which designate them. The names of observers shall be communicated to every other Contracting Party having the right to designate observers, and like notice shall be given of the termination of their appointment.

2. Each observer designated in accordance with the provisions of paragraph 1 of this Article shall have complete freedom of access at any time to any or all areas of Antarctica.

3. All areas of Antarctica, including all stations, installations and equipment within those areas, and all ships and aircraft at points of discharging or embarking cargoes or personnel in Antarctica, shall be open at all times to inspection by any observers designated in accordance with paragraph 1 of this Article.

4. Aerial observation may be carried out at any time over any or all areas of Antarctica by any of the Contracting Parties having the right to designate observers.

5. Each Contracting Party shall, at the time when the present Treaty enters into force for it, inform the other Contracting Parties, and thereafter shall give them notice in advance, of

(*a*) all expeditions to and within Antarctica, on the part of its ships or nationals, and all expeditions to Antarctica organized in or proceeding from its territory;

(*b*) all stations in Antarctica occupied by its nationals; and

(*c*) any military personnel or equipment intended to be introduced by it into Antarctica subject to the conditions prescribed in paragraph 2 of Article I of the present Treaty.

Article VIII

1. In order to facilitate the exercise of their functions under the present Treaty, and without prejudice to the respective positions of the Contracting Parties relating to jurisdiction over all other persons in Antarctica, observers designated under paragraph 1 of Article VII and scientific personnel exchanged under sub-paragraph 1 (*b*) of Article III of the Treaty, and members of the staffs accompanying any such persons, shall be subject only to the jurisdiction of the Contracting Party of which they are nationals in respect of all acts or omissions occurring while they are in Antarctica for the purpose of exercising their functions.

2. Without prejudice to the provisions of paragraph 1 of this Article, and pending the adoption of measures in pursuance of sub-paragraph 1 (*e*) of Article IX, the Contracting Parties concerned in any case of dispute with regard to the exercise of jurisdiction in Antarctica shall immediately consult together with a view to reaching a mutually acceptable solution.

ARTICLE IX

1. Representatives of the Contracting Parties named in the preamble to the present Treaty shall meet at the City of Canberra within two months after the date of entry into force of the Treaty, and thereafter at suitable intervals and places, for the purpose of exchanging information, consulting together on matters of common interest pertaining to Antarctica, and formulating and considering, and recommending to their Governments, measures in furtherance of the principles and objectives of the Treaty, including measures regarding:

(*a*) use of Antarctica for peaceful purposes only;

(*b*) facilitation of scientific research in Antarctica;

(*c*) facilitation of international scientific co-operation in Antarctica;

(*d*) facilitation of the exercise of the rights of inspection provided for in Article VII of the Treaty;

(*e*) questions relating to the exercise of jurisdiction in Antarctica;

(*f*) preservation and conservation of living resources in Antarctica.

2. Each Contracting Party which has become a party to the present Treaty by accession under Article XIII shall be entitled to appoint representatives to participate in the meetings referred to in paragraph 1 of the present Article, during such time as that Contracting Party demonstrates its interest in Antarctica by conducting substantial scientific research activity there, such as the establishment of a scientific station or the despatch of a scientific expedition.

3. Reports from the observers referred to in Article VII of the present Treaty shall be transmitted to the representatives of the Contracting Parties participating in the meetings referred to in paragraph 1 of the present Article.

4. The measures referred to in paragraph 1 of this Article shall become effective when approved by all the Contracting Parties whose representatives were entitled to participate in the meetings held to consider those measures.

5. Any or all of the rights established in the present Treaty may be exercised as from the date of entry into force of the Treaty whether or not any measures facilitating the exercise of such rights have been proposed, considered or approved as provided in this Article.

ARTICLE X

Each of the Contracting Parties undertakes to exert appropriate efforts, consistent with the Charter of the United Nations, to the end that no one engages in any activity in Antarctica contrary to the principles or purposes of the present Treaty.

ARTICLE XI

1. If any dispute arises between two or more of the Contracting Parties concerning the interpretation or application of the present Treaty, those Contracting Parties shall consult among themselves with a view to having the dispute resolved by negotiation,

inquiry, mediation, conciliation, arbitration, judicial settlement or other peaceful means of their own choice.

2. Any dispute of this character not so resolved shall, with the consent, in each case, of all parties to the dispute, be referred to the International Court of Justice for settlement; but failure to reach agreement on reference to the International Court shall not absolve parties to the dispute from the responsibility of continuing to seek to resolve it by any of the various peaceful means referred to in paragraph 1 of this Article.

Article XII

1.—(*a*) The present Treaty may be modified or amended at any time by unanimous agreement of the Contracting Parties whose representatives are entitled to participate in the meetings provided for under Article IX. Any such modification or amendment shall enter into force when the depositary Government has received notice from all such Contracting Parties that they have ratified it.

(*b*) Such modification or amendment shall thereafter enter into force as to any other Contracting Party when notice of ratification by it has been received by the depositary Government. Any such Contracting Party from which no notice of ratification is received within a period of two years from the date of entry into force of the modification or amendment in accordance with the provisions of sub-paragraph 1 (*a*) of this Article shall be deemed to have withdrawn from the present Treaty on the date of the expiration of such period.

2.—(*a*) If after the expiration of thirty years from the date of entry into force of the present Treaty, any of the Contracting Parties whose representatives are entitled to participate in the meetings provided for under Article IX so requests by a communication addressed to the depositary Government, a Conference of all the Contracting Parties shall be held as soon as practicable to review the operation of the Treaty.

(*b*) Any modification or amendment to the present Treaty which is approved at such a Conference by a majority of the Contracting Parties there represented, including a majority of those whose representatives are entitled to participate in the meetings provided for under Article IX, shall be communicated by the depositary Government to all the Contracting Parties immediately after the termination of the Conference and shall enter into force in accordance with the provisions of paragraph 1 of the present Article.

(*c*) If any such modification or amendment has not entered into force in accordance with the provisions of sub-paragraph 1 (*a*) of this Article within a period of two years after the date of its communication to all the Contracting Parties, any Contracting Party may at any time after the expiration of that period give notice to the depositary Government of its withdrawal from the present Treaty; and such withdrawal shall take effect two years after the receipt of the notice by the depositary Government.

Article XIII

1. The present Treaty shall be subject to ratification by the signatory States. It shall be open for accession by any State which is a Member of the United Nations, or by any other State which may be invited to accede to the Treaty with the consent of all the Contracting Parties whose representatives are entitled to participate in the meetings

provided for under Article IX of the Treaty.

2. Ratification of or accession to the present Treaty shall be effected by each State in accordance with its constitutional processes.

3. Instruments of ratification and instruments of accession shall be deposited with the Government of the United States of America, hereby designated as the depositary Government.

4. The depositary Government shall inform all signatory and acceding States of the date of each deposit of an instrument of ratification or accession, and the date of entry into force of the Treaty and of any modification or amendment thereto.

5. Upon the deposit of instruments of ratification by all the signatory States, the present Treaty shall enter into force for those States and for States which have deposited instruments of accession. Thereafter the Treaty shall enter into force for any acceding State upon the deposit of its instrument of accession.

6. The present Treaty shall be registered by the depositary Government pursuant to Article 102 of the Charter of the United Nations.

Article XIV

The present Treaty, done in the English, French, Russian and Spanish languages, each version being equally authentic, shall be deposited in the archives of the Government of the United States of America, which shall transmit duly certified copies thereof to the Government of the signatory and acceding States.

Appendix II
Convention on the Conservation of Antarctic Marine Living Resources

The Contracting Parties,

Recognising the importance of safeguarding the environment and protecting the integrity of the ecosystem of the seas surrounding Antarctica;

Noting the concentration of marine living resources found in Antarctic waters and the increased interest in the possibilities offered by the utilization of these resources as a source of protein;

Conscious of the urgency of ensuring the conservation of Antarctic marine living resources;

Considering that it is essential to increase knowledge of the Antarctic marine ecosystem and its components so as to be able to base decisions on harvesting on sound scientific information;

Believing that the conservation of Antarctic marine living resources calls for international co-operation with due regard for the provisions of the Antarctic Treaty and with the active involvement of all States engaged in research or harvesting activities in Antarctic waters;

Recognising the prime responsibilities of the Antarctic Treaty Consultative Parties for the protection and preservation of the Antarctic environment and, in particular, their responsibilities under Article IX, paragraph 1 (*f*) of the Antarctic Treaty in respect of the preservation and conservation of living resources in Antarctica;

Recalling the action already taken by the Antarctic Treaty Consultative Parties including in particular the Agreed Measures for the Conservation of Antarctic Fauna and Flora, as well as the provisions of the Convention for the Conservation of Antarctic Seals;

Bearing in mind the concern regarding the conservation of Antarctic marine living resources expressed by the Consultative Parties at the Ninth Consultative Meeting of the Antarctic Treaty and the importance of the provisions of Recommendation IX-2 which led to the establishment of the present Convention;

Believing that it is in the interest of all mankind to preserve the waters surrounding the Antarctic continent for peaceful purposes only and to prevent their becoming the scene or object of international discord;

Recognising, in the light of the foregoing, that it is desirable to establish suitable machinery for recommending, promoting, deciding upon and co-ordinating the measures and scientific studies needed to ensure the conservation of Antarctic marine living organisms;

Have agreed as follows:

ARTICLE 1

1. This Convention applies to the Antarctic marine living resources of the area south of 60° South latitude and to the Antarctic marine living resources of the area between that latitude and the Antarctic Convergence which form part of the Antarctic marine ecosystem.

2. Antarctic marine living resources means the populations of fin fish, molluscs, crustaceans and all other species of living organisms, including birds, found south of the Antarctic Convergence.

3. The Antarctic marine ecosystem means the complex of relationships of Antarctic marine living resources with each other and with their physical environment.

4. The Antarctic Convergence shall be deemed to be a line joining the following points along parallels of latitude and meridians of longitude: 50°S, O°; 50°S, 30°E; 45°S, 30°E; 45°S, 80°E; 55°S, 80°E; 55°S, 150°E; 60°S, 150°E; 60°S, 50°W; 50°S, 50°W; 50°S, 0°.

ARTICLE II

1. The objective of this Convention is the conservation of Antarctic marine living resources.

2. For the purposes of this Convention, the term 'conservation' includes rational use.

3. Any harvesting and associated activities in the area to which this Convention applies shall be conducted in accordance with the provisions of this Convention and with the following principles of conservation:

(*a*) prevention of decrease in the size of any harvested population to levels below those which ensure its stable recruitment. For this purpose its size should not be allowed to fall below a level close to that which ensures the greatest net annual increment;

(*b*) maintenance of the ecological relationships between harvested, dependent and related populations of Antarctic marine living resources and the restoration of depleted populations to the levels defined in sub-paragraph (*a*) above; and

(*c*) prevention of changes or minimization of the risk of changes in the marine ecosystem which are not potentially reversible over two or three decades, taking into account the state of available knowledge of the direct and indirect impact of harvesting, the effect of the introduction of alien species, the effects of associated activities on the marine ecosystem and of the effects of environmental changes, with the aim of making possible the sustained conservation of Antarctic marine living resources.

ARTICLE III

The Contracting Parties, whether or not they are Parties to the Antarctic Treaty, agree that they will not engage in any activities in the Antarctic Treaty area contrary to the principles and purposes of that Treaty and that, in their relations with each other, they are bound by the obligations contained in Articles I and V of the Antarctic Treaty.

Article IV

1. With respect to the Antarctic Treaty area, all Contracting Parties, whether or not they are Parties to the Antarctic Treaty, are bound by Articles IV and VI of the Antarctic Treaty in their relations with each other.

2. Nothing in this Convention and no acts or activities taking place while the present Convention is in force shall:

(*a*) constitute a basis for asserting, supporting or denying a claim to territorial sovereignty in the Antarctic Treaty area or create any rights of sovereignty in the Antarctic Treaty area;

(*b*) be interpreted as a renunciation or diminution by any Contracting Party of, or as prejudicing, any right or claim or basis of claim to exercise coastal state jurisdiction under international law within the area to which this Convention applies;

(*c*) be interpreted as prejudicing the position of any Contracting Party as regards its recognition or non-recognition of any such right, claim or basis of claim;

(*d*) affect the provision of Article IV, paragraph 2, of the Antarctic Treaty that no new claim, or enlargement of an existing claim, to territorial sovereignty in Antarctica shall be asserted while the Antarctic Treaty is in force.

Article V

1. The Contracting Parties which are not Parties to the Antarctic Treaty acknowledge the special obligations and responsibilities of the Antarctic Treaty Consultative Parties for the protection and preservation of the environment of the Antarctic Treaty area.

2. The Contracting Parties which are not Parties to the Antarctic Treaty agree that, in their activities in the Antarctic Treaty area, they will observe as and when appropriate the Agreed Measures for the Conservation of Antarctic Fauna and Flora and such other measures as have been recommended by the Antarctic Treaty Consultative Parties in fulfilment of their responsibility for the protection of the Antarctic environment from all forms of harmful human interference.

3. For the purposes of this Convention, 'Antarctic Treaty Consultative Parties' means the Contracting Parties to the Antarctic Treaty whose representatives participate in meetings under Article IX of the Antarctic Treaty.

Article VI

Nothing in this Convention shall derogate from the rights and obligations of Contracting Parties under the International Convention for the Regulation of Whaling and the Convention for the Conservation of Antarctic Seals.

ARTICLE VII

1. The Contracting Parties hereby establish and agree to maintain the Commission for the Conservation of Antarctic Marine Living Resources (hereinafter referred to as 'the Commission').

2. Membership in the Commission shall be as follows:

(*a*) each Contracting Party which participated in the meeting at which this Convention was adopted shall be a Member of the Commission;

(*b*) each State Party which has acceded to this Convention pursuant to Article XXIX shall be entitled to be a Member of the Commission during such time as that acceding party is engaged in research or harvesting activities in relation to the marine living resources to which this Convention applies;

(*c*) each regional economic integration organization which has acceded to this Convention pursuant to Article XXIX shall be entitled to be a Member of the Commission during such time as its States members are so entitled;

(*d*) a Contracting Party seeking to participate in the work of the Commission pursuant to sub-paragraphs (*b*) and (*c*) above shall notify the Depositary of the basis upon which it seeks to become a Member of the Commission and of its willingness to accept conservation measures in force. The Depositary shall communicate to each Member of the Commission such notification and accompanying information. Within two months of receipt of such communication from the Depositary, any Member of the Commission may request that a special meeting of the Commission be held to consider the matter. Upon receipt of such request, the Depositary shall call such a meeting. If there is no request for a meeting, the Contracting Party submitting the notification shall be deemed to have satisfied the requirements for Commission Membership.

3. Each Member of the Commission shall be represented by one representative who may be accompanied by alternate representatives and advisers.

ARTICLE VIII

The Commission shall have legal personality and shall enjoy in the territory of each of the States Parties such legal capacity as may be necessary to perform its function and achieve the purposes of this Convention. The privileges and immunities to be enjoyed by the Commission and its staff in the territory of a State Party shall be determined by agreement between the Commission and the State Party concerned.

ARTICLE IX

1. The function of the Commission shall be to give effect to the objective and principles set out in Article II of this Convention. To this end, it shall:

(*a*) facilitate research into and comprehensive studies of Antarctic marine living resources and of the Antarctic marine ecosystem;

(*b*) compile data on the status of and changes in population of Antarctic marine living resources and on factors affecting the distribution, abundance and productivity of harvested species and dependent or related species or populations;

(*c*) ensure the acquisition of catch and effort statistics on harvested populations;

(*d*) analyse, disseminate and publish the information referred to in sub-paragraphs (*b*) and (*c*) above and the reports of the Scientific Committee;

(*e*) identify conservation needs and analyse the effectiveness of conservation measures;

(*f*) formulate, adopt and revise conservation measures on the basis of the best scientific evidence available, subject to the provisions of paragraph 5 of this Article;

(*g*) implement the system of observation and inspection established under Article XXIV of this Convention;

(*h*) carry out such other activities as are necessary to fulfil the objectives of this Convention.

2. The conservation measures referred to in paragraph 1 (*f*) above include the following:

(*a*) the designation of the quantity of any species which may be harvested in the area to which this Convention applies;

(*b*) the designation of regions and sub-regions based on the distribution of populations of Antarctic marine living resources;

(*c*) the designation of the quantity which may be harvested from the populations of regions and sub-regions;

(*d*) the designation of protected species;

(*e*) the designation of the size, age and, as appropriate, sex of species which may be harvested;

(*f*) the designation of open and closed seasons for harvesting;

(*g*) the designation of the opening and closing of areas, regions or sub-regions for purposes of scientific study or conservation, including special areas for protection and scientific study;

(*h*) regulation of the effort employed and methods of harvesting, including fishing gear, with a view, *inter alia*, to avoiding undue concentration of harvesting in any region or sub-region;

(*i*) the taking of such other conservation measures as the Commission considers necessary for the fulfilment of the objective of this Convention, including measures concerning the effects of harvesting and associated activities on components of the marine ecosystem other than the harvested populations.

3. The Commission shall publish and maintain a record of all conservation measures in force.

4. In exercising its functions under paragraph 1 above, the Commission shall take full account of the recommendations and advice of the Scientific Committee.

5. The Commission shall take full account of any relevant measures or regulations established or recommended by the Consultative Meetings pursuant to Article IX of the Antarctic Treaty or by existing fisheries commissions responsible for species which may enter the area to which this Convention applies, in order that there shall be no inconsistency between the rights and obligations of a Contracting Party under such

regulations or measures and conservation measures which may be adopted by the Commission.

6. Conservation measures adopted by the Commission in accordance with this Convention shall be implemented by Members of the Commission in the following manner.

(*a*) the Commission shall notify conservation measures to all Members of the Commission;

(*b*) conservation measures shall become binding upon all Members of the Commission 180 days after such notification, except as provided in sub-paragraphs (*c*) and (*d*) below;

(*c*) if a Member of the Commission, within ninety days following the notification specified in sub-paragraph (*a*), notifies the Commission that it is unable to accept the conservation measure, in whole or in part, the measure shall not, to the extent stated, be binding upon that Member of the Commission;

(*d*) in the event that any Member of the Commission invokes the procedure set forth in sub-paragraph (*c*) above, the Commission shall meet at the request of any Member of the Commission to review the conservation measure. At the time of such meeting and within thirty days following the meeting, any member of the Commission shall have the right to declare that it is no longer able to accept the conservation measure, in which case the Member shall no longer be bound by such measure.

ARTICLE X

1. The Commission shall draw the attention of any State which is not a Party to this Convention to any activity undertaken by its nationals or vessels which, in the opinion of the Commission, affects the implementation of the objectives of this Convention.

2. The Commission shall draw the attention of all Contracting Parties to any activity which, in the opinion of the Commission, affects the implementation by a Contracting Party of the objective of this Convention or the compliance by that Contracting Party with its obligations under this Convention.

ARTICLE XI

The Commission shall seek to cooperate with Contracting Parties which may exercise jurisdiction in marine areas adjacent to the area to which this Convention applies in respect of the conservation of any stock or stocks of associated species which occur both within those areas and the area to which this Convention applies, with a view to harmonizing the conservation measures adopted in respect of such stocks.

ARTICLE XII

1. Decisions of the Commission on matters of substance shall be taken by consensus. The question of whether a matter is one of substance shall be treated as a matter of substance.

2. Decisions on matters other than those referred to in paragraph 1 above shall be

taken by a simple majority of the Members of the Commission present and voting.

3. In Commission consideration of any item requiring a decision, it shall be made clear whether a regional economic integration organization will participate in the taking of the decision and, if so, whether any of its member States will also participate. The number of Contracting Parties so participating shall not exceed the number of member States of the regional economic integration organization which are Members of the Commission.

4. In the taking of decisions pursuant to this Article, a regional economic integration organization shall have only one vote.

Article XIII

1. The headquarters of the Commission shall be established at Hobart, Tasmania, Australia.

2. The Commission shall hold a regular annual meeting. Other meetings shall also be held at the request of one-third of its members and as otherwise provided in this Convention. The first meeting of the Commission shall be held within three months of the entry into force of this Convention, provided that among the Contracting Parties there are at least two States conducting harvesting activities within the area to which this Convention applies. The first meeting shall, in any event, be held within one year of the entry into force of this Convention. The Depositary shall consult with the signatory States regarding the first Commission meeting, taking into account that a broad representation of such States is necessary for the effective operation of the Commission.

3. The Depositary shall convene the first meeting of the Commission at the headquarters of the Commission. Thereafter, meetings of the Commission shall be held at its headquarters, unless it decides otherwise.

4. The Commission shall elect from among its members a Chairman and Vice-Chairman, each of whom shall serve for a term of two years and shall be eligible for re-election for one additional term. The first Chairman shall, however, be elected for an initial term of three years. The Chairman and Vice-Chairman shall not be representatives of the same Contracting Party.

5. The Commission shall adopt and amend as necessary the rules of procedure for the conduct of its meetings, except with respect to the matters dealt with in Article XII of this Convention.

6. The Commission may establish such subsidiary bodies as are necessary for the performance of its functions.

Article XIV

1. The Contracting Parties hereby establish the Scientific Committee for the Conservation of Antarctic Marine Living Resources (hereinafter referred to as 'the Scientific Committee') which shall be a consultative body to the Commission. The Scientific Committee shall normally meet at the headquarters of the Commission unless the Scientific Committee decides otherwise.

2. Each Member of the Commission shall be a member of the Scientific Committee and shall appoint a representative with suitable scientific qualifications who may be accompanied by other experts and advisers.

3. The Scientific Committee may seek the advice of other scientists and experts as may be required on an *ad hoc* basis.

Article XV

1. The Scientific Committee shall provide a forum for consultation and cooperation concerning the collection, study and exchange of information with respect to the marine living resources to which this Convention applies. It shall encourage and promote cooperation in the field of scientific research in order to extend knowledge of the marine living resources of the Antarctic marine ecosystem.

2. The Scientific Committee shall conduct such activities as the Commission may direct in pursuance of the objective of this Convention and shall:

(*a*) establish criteria and methods to be used for determinations concerning the conservation measures referred to in Article IX of this Convention;

(*b*) regularly assess the status and trends of the populations of Antarctic marine living resources;

(*c*) analyse data concerning the direct and indirect effects of harvesting on the populations of Antarctic marine living resources;

(*d*) assess the effects of proposed changes in the methods or levels of harvesting and proposed conservation measures;

(*e*) transmit assessments, analyses, reports and recommendations to the Commission as requested or on its own initiative regarding measures and research to implement the objective of this Convention;

(*f*) formulate proposals for the conduct of international and national programs of research into Antarctic marine living resources.

3. In carrying out its functions, the Scientific Committee shall have regard to the work of other relevant technical and scientific organizations and to the scientific activities conducted within the framework of the Antarctic Treaty.

Article XVI

1. The first meeting of the Scientific Committee shall be held within three months of the first meeting of the Commission. The Scientific Committee shall meet thereafter as often as may be necessary to fulfil its functions.

2. The Scientific Committee shall adopt and amend as necessary its rules of procedure. The rules and any amendments thereto shall be approved by the Commission. The rules shall include procedures for the presentation of minority reports.

3. The Scientific Committee may establish, with the approval of the Commission, such subsidiary bodies as are necessary for the performance of its functions.

ARTICLE XVII

1. The Commission shall appoint an Executive Secretary to serve the Commission and Scientific Committee according to such procedures and on such terms and conditions as the Commission may determine. His term of office shall be for four years and he shall be eligible for re-appointment.

2. The Commission shall authorize such staff establishment for the Secretariat as may be necessary and the Executive Secretary shall appoint, direct and supervise such staff according to such rules and procedures and on such terms and conditions as the Commission may determine.

3. The Executive Secretary and Secretariat shall perform the functions entrusted to them by the Commission.

ARTICLE XVIII

The official languages of the Commission and of the Scientific Committee shall be English, French, Russian and Spanish.

ARTICLE XIX

1. At each annual meeting, the Commission shall adopt by consensus its budget and the budget of the Scientific Committee.

2. A draft budget for the Commission and the Scientific Committee and any subsidiary bodies shall be prepared by the Executive Secretary and submitted to the Members of the Commission at least sixty days before the annual meeting of the Commission.

3. Each Member of the Commission shall contribute to the budget. Until the expiration of five years after the entry into force of this Convention, the contribution of each Member of the Commission shall be equal. Thereafter the contribution shall be determined in accordance with two criteria: the amount harvested and an equal sharing among all Members of the Commission. The Commission shall determine by consensus the proportion in which these two criteria shall apply.

4. The financial activities of the Commission and Scientific Committee shall be conducted in accordance with financial regulations adopted by the Commission and shall be subject to an annual audit by external auditors selected by the Commission.

5. Each Member of the Commission shall meet its own expenses arising from attendance at meetings of the Commission and of the Scientific Committee.

6. A Member of the Commission that fails to pay its contributions for two consecutive years shall not, during the period of its default, have the right to participate in the taking of decisions in the Commission.

ARTICLE XX

1. The Members of the Commission shall, to the greatest extent possible, provide annually to the Commission and to the Scientific Committee such statistical, biological

and other data and information as the Commission and Scientific Committee may require in the exercise of their functions.

2. The Members of the Commission shall provide, in the manner and at such intervals as may be prescribed, information about their harvesting activities, including fishing areas and vessels, so as to enable reliable catch and effort statistics to be compiled.

3. The Members of the Commission shall provide to the Commission at such intervals as may be prescribed information on steps taken to implement the conservation measures adopted by the Commission.

4. The Members of the Commission agree that in any of their harvesting activities, advantage shall be taken of opportunities to collect data needed to assess the impact of harvesting.

Article XXI

1. Each Contracting Party shall take appropriate measures within its competence to ensure compliance with the provisions of this Convention and with conservation measures adopted by the Commission to which the Party is bound in accordance with Article IX of this Convention.

2. Each Contracting Party shall transmit to the Commission information on measures taken pursuant to paragraph 1 above, including the imposition of sanctions for any violation.

Article XXII

1. Each Contracting Party undertakes to exert appropriate efforts, consistent with the Charter of the United Nations, to the end that no one engages in any activity contrary to the objective of this Convention.

2. Each Contracting Party shall notify the Commission of any such activity which comes to its attention.

Article XXIII

1. The Commission and the Scientific Committee shall co-operate with the Antarctic Treaty Consultative Parties on matters falling within the competence of the latter.

2. The Commission and the Scientific Committee shall co-operate, as appropriate, with the Food and Agriculture Organisation of the United Nations and with other Specialised Agencies.

3. The Commission and the Scientific Committee shall seek to develop co-operative working relationships, as appropriate, with inter-governmental and non-governmental organizations which could contribute to their work including the Scientific Committee on Antarctic Research, the Scientific Committee on Oceanic Research and the International Whaling Commission.

4. The Commission may enter into agreements with the organizations referred to in this Article and with other organizations as may be appropriate. The Commission and the Scientific Committee may invite such organizations to send observers to their meetings and to meetings of their subsidiary bodies.

Article XXIV

1. In order to promote the objective and ensure observance of the provisions of this Convention, the Contracting Parties agree that a system of observation and inspection shall be established.

2. The system of observation and inspection shall be elaborated by the Commission on the basis of the following principles:

(*a*) Contracting Parties shall co-operate with each other to ensure the effective implementation of the system of observation and inspection, taking account of the existing international practice. This system shall include, *inter alia*, procedures for boarding and inspection by observers and inspectors designated by the Members of the Commission and procedures for flag state prosecution and sanctions on the basis of evidence resulting from such boarding and inspections. A report of such prosecutions and sanctions imposed shall be included in the information referred to in Article XXI of this Convention;

(*b*) in order to verify compliance with measures adopted under this Convention, observation and inspection shall be carried out on board vessels engaged in scientific research or harvesting of marine living resources in the area to which this Convention applies, through observers and inspectors designated by the Members of the Commission and operating under terms and conditions to be established by the Commission;

(*c*) designated observers and inspectors shall remain subject to the jurisdiction of the Contracting Party of which they are nationals. They shall report to the Member of the Commission by which they have been designated which in turn shall report to the Commission.

3. Pending the establishment of the system of observation and inspection, the Members of the Commission shall seek to establish interim arrangements to designate observers and inspectors and such designated observers and inspectors shall be entitled to carry out inspections in accordance with the principles set out in paragraph 2 above.

Article XXV

1. If any dispute arises between two or more of the Contracting Parties concerning the interpretation or application of this Convention, those Contracting Parties shall consult among themselves with a view to having the dispute resolved by negotiation, inquiry, mediation, conciliation, arbitration, judicial settlement or other peaceful means of their own choice.

2. Any dispute of this character not so resolved shall, with the consent in each case of all Parties to the dispute, be referred for settlement to the International Court of Justice or to arbitration; but failure to reach agreement on reference to the International Court or to arbitration shall not absolve Parties to the dispute from the responsibility of continuing to seek to resolve it by any of the various peaceful means referred to in paragraph 1 above.

3. In cases where the dispute is referred to arbitration, the arbitral tribunal shall be constituted as provided in the Annex to this Convention.

Article XXVI

1. This Convention shall be open for signature at Canberra from 1 August to 31 December 1980 by the States participating in the Conference on the Conservation of Antarctic Marine Living Resources held at Canberra from 7 to 20 May 1980.

2. The States which so sign will be the original signatory States of the Convention.

Article XXVII

1. This Convention is subject to ratification, acceptance or approval by signatory States.

2. Instruments of ratification, acceptance or approval shall be deposited with the Government of Australia, hereby designated as the Depositary.

Article XXVIII

1. This Convention shall enter into force on the thirtieth day following the date of deposit of the eighth instrument of ratification, acceptance or approval by States referred to in paragraph 1 of Article XXVI of this Convention.

2. With respect to each State or regional economic integration organization which subsequent to the date of entry into force of this Convention deposits an instrument of ratification, acceptance, approval or accession, the Convention shall enter into force on the thirtieth day following such deposit.

Article XXIX

1. This Convention shall be open for accession by any State interested in research or harvesting activities in relation to the marine living resources to which this Convention applies.

2. This Convention shall be open for accession by regional economic integration organizations constituted by sovereign States which include among their members one or more States Members of the Commission and to which the States Members of the organization have transferred, in whole or in part, competences with regard to the matters covered by this Convention. The accession of such regional economic integration organizations shall be the subject of consultations among Members of the Commission.

Article XXX

1. This Convention may be amended at any time.

2. If one-third of the Members of the Commission request a meeting to discuss a proposed amendment the Depositary shall call such a meeting.

3. An amendment shall enter into force when the Depositary has received instruments of ratification, acceptance or approval thereof from all the Members of the Commission.

4. Such amendment shall thereafter enter into force as to any other Contracting Party when notice of ratification, acceptance or approval by it has been received by the Depositary. Any such Contracting Party from which no such notice has been received within a period of one year from the date of entry into force of the amendment in accordance with paragraph 3 above shall be deemed to have withdrawn from this Convention.

Article XXXI

1. Any Contracting Party may withdraw from this Convention on 30 June of any year, by giving written notice not later than 1 January of the same year to the Depositary, which, upon receipt of such a notice, shall communicate it forthwith to the other Contracting Parties.

2. Any other Contracting Party may, within sixty days of the receipt of a copy of such a notice from the Depositary, give written notice of withdrawal to the Depository in which case the Convention shall cease to be in force on 30 June of the same year with respect to the Contracting Party giving such notice.

3. Withdrawal from this Convention by any Member of the Commission shall not affect its financial obligations under this Convention.

Article XXXII

The Depositary shall notify all Contracting Parties of the following:

(*a*) signatures of this Convention and the deposit of instruments of ratification, acceptance, approval or accession;

(*b*) the date or entry into force of this Convention and of any amendment thereto.

Article XXXIII

1. This Convention, of which the English, French, Russian and Spanish texts are equally authentic, shall be deposited with the Government of Australia which shall transmit duly certified copies thereof to all signatory and acceding Parties.

2. This Convention shall be registered by the Depositary pursuant to Article 102 of the Charter of the United Nations.

Drawn up at Canberra this twentieth day of May 1980.

In witness whereof the undersigned, being duly authorized, have signed this Convention.

Annex for an Arbitral Tribunal

1. The arbitral tribunal referred to in paragraph 3 of Article XXV shall be composed of three arbitrators who shall be appointed as follows:

(*a*) The Party commencing proceedings shall communicate the name of an arbitrator to the other Party which, in turn, within a period of forty days following such notification, shall communicate the name of the second arbitrator. The Parties shall, within a period of sixty days following the appointment of the second arbitrator, appoint the third arbitrator, who shall not be a national of either Party and shall not be of the same nationality as either of the first two arbitrators. The third arbitrator shall preside over the tribunal.

(*b*) If the second arbitrator has not been appointed within the prescribed period, or if the Parties have not reached agreement within the prescribed period on the appointment of the third arbitrator, that arbitrator shall be appointed, at the request of either Party, by the Secretary-General of the Permanent Court of Arbitration, from among persons of international standing not having the nationality of a State which is a Party to this Convention.

2. The arbitral tribunal shall decide where its headquarters will be located and shall adopt its own rules of procedure.

3. The award of the arbitral tribunal shall be made by a majority of its members, who may not abstain from voting.

4. Any Contracting Party which is not a Party to the dispute may intervene in the proceedings with the consent of the arbitral tribunal.

5. The award of the arbitral tribunal shall be final and binding on all Parties to the dispute and on any Party which intervenes in the proceedings and shall be complied with without delay. The arbitral tribunal shall interpret the award at the request of one of the Parties to the dispute or of any intervening Party.

6. Unless the arbitral tribunal determines otherwise because of the particular circumstances of the case, the expenses of the tribunal, including the remuneration of its members, shall be borne by the Parties to the dispute in equal shares.

Appendix III

Extract from the Final Act of the Conference on the Conservation of Antarctic Marine Living Resources Canberra, 7–20 May 1980

The Conference also decided to include in the Final Act the text of the following statement made by the Chairman on 19 May 1980 regarding the application of the Convention on the Conservation of Antarctic Marine Living Resources to the waters adjacent to Kerguelen and Crozet over which France has jurisdiction and to waters adjacent to other islands within the area to which this Convention applies over which the existence of State sovereignty is recognized by all Contracting Parties.

'1. Measures for the conservation of Antarctic marine living resources of the waters adjacent to Kerguelen and Crozet, over which France has jurisdiction, adopted by France prior to the entry into force of the Convention, would remain in force after the entry into force of the Convention until modified by France acting within the framework of the Commission or otherwise.

2. After the Convention has come into force, each time the Commission should undertake examination of the conservation needs of the marine living resources of the general area in which the waters adjacent to Kerguelen and Crozet are to be found, it would be open to France either to agree that the waters in question should be included in the area of application of any specific conservation measure under consideration or to indicate that they should be excluded. In the latter event, the Commission would not proceed to the adoption of the specific conservation measure in a form applicable to the waters in question unless France removed its objection to it. France could also adopt such national measures as it might deem appropriate for the waters in question.

3. Accordingly, when specific conservation measures are considered within the framework of the Commission and with the participation of France, then:

(*a*) France would be bound by any conservation measures adopted by consensus with its participation for the duration of those measures. This would not prevent France from promulgating national measures that were more strict than the Commission's measures or which dealt with other matters;

(*b*) In the absence of consensus, France could promulgate any national measures which it might deem appropriate.

4. Conservation measures, whether national measures or measures adopted by the Commission, in respect of the waters adjacent to Kerguelen and Crozet, would be enforced by France. The system of observation and inspection foreseen by the Convention would not be implemented in the waters adjacent to Kerguelen and Crozet except as agreed by France and in the manner so agreed.

5. The understandings, set forth in paragraphs 1–4 above, regarding the application of the Convention to waters adjacent to the Islands of Kerguelen and Crozet, also apply to waters adjacent to the islands within the area to which this Convention applies over which the existence of State sovereignty is recognized by all Contracting Parties.'

No objection to the statement was made.

Index

accretion of territory, 13
Afghanistan, 81
Aga Khan, Prince Sadruddin, 112
Agreed Measures for the Conservation of Antarctic Fauna and Flora, 34, 176
Air New Zealand, 152; crash of plane, 152-4, 152
Alaska: oil spill, 60-1; pipeline, 61
Algeria, 102
Alyeska Pipe Line Service Company, 62
Amerasinghe, Shirley, 72
Amnesty International, 127
Amoco Cadiz, 60
Antarctic Treaty (1959), 7, 9-25, 28, 77, 79-80, 105, 106, 114, 135, 145, 146, 149-58, 173; limitations of, 22-4
Antarctica: as last wilderness, 2; control of, 145; importance of, 1-3; legal status of, 163-5; national claims to, 15-17; ownership of, 145, 172; population of, 15
Antarctic and Southern Oceans Coalition (ASOC), 43, 44, 134, 137
Antarctic Defence Coalition, 72
Antarctic Mineral Resources Commission, 56
Antarctic Public Heritage Agency (APHA), 173, 174, 177, 178
Antarctica Project, 58, 100, 134, 157
anti-uranium movement, 127
Antigua, 75
apartheid, 85, 96
Argentina, 4, 16, 17, 19, 21, 22, 24, 29, 41, 46, 52, 57, 73, 110, 164
Association of South East Asian Nations (ASEAN), 76
astronomy, 2
Auburn, F.M., 40
Australia, 4, 18, 19, 21, 29, 31, 33, 41, 42, 44, 46, 49, 50, 52, 56, 57, 64, 65, 67, 72, 75, 79-80, 81, 82, 83, 86, 87-92, 95, 97, 105, 106, 107, 109, 128, 142, 149, 150, 151, 172, 178, 179; refusal to sign CRAMRA, 59-63
Australian Antarctic Territory (AAT), 87, 88-90, 137, 154
Austria, 46, 52

Bahia Paraiso, 109, 110, 155
Barnes, James, 157
Bates, G.M., 170-1
Beeby, Chris, 52, 65
Beesley, J. Alan, 169
Behar, Abraham, 102
Belgium, 19, 23, 41, 46, 52, 66
Bellingshausen, Thaddeus, 16
Bin Mohamad, Mahathir, 75-7, 80
Binding Triad system, 173, 178
Bird, Vere, 75
Blanco, M., 72-3
Bloxham, Les, 154
Bogart, Paul, 109
Bohlen, Buff, 67
Bolivia, 169
Booker, Malcolm, 91
Brandt, Willy, 112
Brazil, 41, 52, 57, 65, 72, 73, 74, 75, 83, 169
British Petroleum (BP), 67
Brundtland, Gro Harlem, 115
Brundtland Report, 116
Bulgaria, 46, 52
bureaucracy, 59
Bush, George, 67
Button, John, 109
Byrd, Admiral Richard, 20, 164

Callendar, G.S., 121
Campaign for Nuclear Disarmament (CND), 139, 140
Canada, 35, 55
carbon dioxide, 121, 122
Casey, R.G., 87, 88

Convention for the Conservation of Antarctic Marine Living Resources (CCAMLR), 34, 37-9, 98-9, 106, 114, 134, 135, 145, 146, 149, 150, 178, 185, 203-4; assessed, 39-44
Centre for War/Peace Studies, 173
cession of territory, 14
Chadderton, John, 103
Chile, 2, 4, 16, 17, 19, 21, 22, 24, 29, 41, 44, 46, 52, 57, 64, 73, 109, 164; claims on Antarctica, 78-9
China, 52, 79, 82, 92, 110, 121, 140, 177, 179
chlorofluorocarbons (CFCs), 119-20
Clean Air Act (1956) (UK), 123
Club of Rome, 138
coal deposits, 49
cobalt, 71, 167
Cod Wars, 21
cod, Antarctic, 41, 42-4
Cold War, 20, 178
colonialism, 14, 15
Columbus, Christopher, 162
commissions, international, 112-13
common heritage, concept of, 160; limitations of, 165-8
commons, global, 114, 115, 116, 159-60
Community Aid Abroad, 127
conservation, 39, 113-15
Consultative Parties, 1, 23, t27, 40, 46, 51, 52, 74, 87, 92, 97, 100, 101, 107, 147, 149, 170, 176-8
Convention for the Conservation of Atlantic Seals, 32-3, 150
Convention on the Regulation of Antarctic Mineral Resource Activities (CRAMRA), 55-65
Cook, Captain James, 16, 162
copper, 48, 71
Cousteau, Jacques, 63-4
Cuba, 52
customary rule, 11
Cuthbert, Arani, 66
Cyprus, 21
Czechoslovakia, 52

de Klerk, F.W., 99
debt crisis, 141
Delmage, Mark, 102
demilitarization, 79, 91, 117, 172, 176
demilitarized zone, Antarctica as, 21
Denmark, 52
denuclearization, 79
dinosaurs, 3
disarmament, 174, 180
Discovery, 88
disputes, peaceful settlement of, 174
Dolphin Action and Protection Group (DPAG), 99
Dorn, Walter, 175-6
Drake, Sir Francis, 11
driftnet fishing, 30-1, 83
Dumont d'Urville base, 100, 101, 136, 177

East Timor, 86
ECO newspaper, 134
economic development, ban on, 176
Ecuador, 52
environmental movement, 64, 107, 141
environmental protection, 24, 53, 58, 59, 60, 61, 63, 70, 99, 110, 116, 128, 137, 172
Environmental Protection Agency, 157-8
Esperanza base, 110
Ethiopia, 14
European Community (EC), 37, 41
Evans, Gareth, 103
Evatt, H.V., 92
Evensen Group, 72
Exclusive Economic Zone (EEZ), 166
Exxon company, 60-1
Exxon Valdez, 60

Faux, Jeff, 129
Fifth World, 127-42, 179
Finland, 23, 44, 52, 57, 84
fishing industry, 1, 24, 35, 47, 70, 71, 135, 180
Fitzmaurice, Peter, 155
Ford company, 129
fossils, 3, 47
France, 4, 19, 21, 37, 41, 46, 52, 57, 59, 65, 66, 67, 91, 92, 95, 99, 140, 149, 150, 176, 179; opposition to CRAMRA, 63-5
Fraser, Malcolm, 31, 90

Gandhi, Indira, 74
Gauss, 17
genetic diversity, preservation of, 113
Genocide Convention, 12
Gentle, Tim, 33
Germany, 29, 178; East, 22, 37, 41, 42, 52; West, 37, 41, 52, 75
Giscard d'Estaing, Valery, 175
Glomar Challenger, 49
Goldsworthy, Lyn, 44
Gondwana, 134
Gondwanaland, 47, 48
Gorbachev, M.S., 179
Greece, 46, 52
Greene, Harold, 153
greenhouse effect, 3, 5, 63, 67, 95, 121-5, 141, 144, 169

Greenpeace, 33, 66, 100, 101, 102, 108, 134-5, 136
Grotius, Hugo, 161
Groves, Brenton, 105

habitat loss, 100
Hague Conference, 123, 125
Hanse, James, 141
Hardin, Garrett, 159
Hassan bin Talal, Prince, 112
Hawke, Bob, 6, 59, 62
Hayden, Bill, 79
Hazelwood, Captain, 60, 62
Helmut Rhode company, 154
Heritage Foundation, 167
Hernu, Charles, 104
Hollings, Les, 61
Howard, John, 63
Howe, Geoffrey, 75
Hudson, Richard, 173-4
Huidobro, Hernan, 78
Hungary, 46, 52

ice sheet of Antarctica, 4, 49, 122; breaking up of, 123; melting of, 95 (effect of, 6)
ice, harvesting of, 155-7
icebergs, 50, 54
Iceland, 30
IMBS company, 156
independence of colonies, 15
India, 23, 41, 46, 52, 73-5, 79, 82, 110, 121
Indonesia, 86
Intermediate Nuclear Forces Agreement (INFA), 21
International Centre for Ocean Development, 35
International Commission of Jurists, 127
International Committee of the Red Cross (ICRC), 133
International Convention for the Regulation of Whaling, 38
International Court of Justice (ICJ), 10, 12, 17, 20, 102, 124, 165
International Geophysical Year (IGY), 2, 9, 17-19, 23, 74
International Monitoring Agency for the Arctic, 175
International Seabed Authority, 137, 166-7
International Union for Conservation of Nature (IUCN), 113
International Whaling Commission (IWC), 28, 29, 149
internationalization of Antarctica, 90, 91
Iran, 49
iron deposits, 49

Italy, 14, 46, 52, 66
Ivanov, Vladimir, 33
Ixtoc well blowout, 62

Japan, 19, 29, 31, 32, 33, 41, 44, 47, 50, 52, 162, 163, 178

Kerguelen Islands, 37, 42, 100, 102-6
King, H.G.R., 17
Korea, South, 84
krill, 3, 40, 51, 54, 136
krill fishing, 34, 36-7, 43, 44, 47, 48, 89, 114, 135, 145
Kuper, Leo, 12
Kurtz, Harriet, 174
Kurtz, Howard, 174

Labor Party (Australia), 89, 92
Labour Party (New Zealand), 144
Labour Party (UK), 138
Law of the Sea Conference, 69-72, 162, 166
Law of the Sea Treaty, 59, 167
law, international, 9-13
Ledun, Bernard, 102
Lemkin, Raphael, 12
Liberal Party (Australia), 90, 92
Linblad, Lars Eric, 150
Lovelock, James, 119
Lupinacci, Julio Cesar, 81
Lutz, Robert, 129

Mahon, Peter, 153
Malaysia, 70, 75-7, 78, 80, 82, 83, 84, 86, 92, 96, 144, 146, 147, 148, 149, 177
Malta, 69, 77, 163, 166
Malvinas/Falklands dispute, 17, 22, 146
manganese, 71, 166
Manheim, Bruce, 110
Manley, Michael, 112
Marie Byrd Land, 20, 21
marine technology, 70
Marion Island, 98-9
market system, 160
Marshall, Russell, 57-8
Mawson, Douglas, 89
McGuinness, Padraic, 64, 107
McMurdo base, 108, 110, 151, 152, 153
Mead, Margaret, 138
meteorites, 54
Mexico, 62
Minerals Act (UK), 66
minerals, in Antarctica, 46-67
mining in Antarctia, 6, 49, 54; ban on, 67, 85; regulation of, 52, 58
mining, sea-bed, 166, 167, 168
Minority Rights Group, 127

Mitchell, Barbara, 50
Molina, Mario, 119
Montreal Protocol on Ozone Layer, 120-1
Moore, Margaret, 58-9
Muldoon, Robert, 153
Mururoa, 102, 103, 104

Namibia, 86
National Aeronautics and Space Administration (NASA), 175
national liberation movements, 14
National Science Foundation (US), 108-9
Netherlands, 30, 52
New International Economic Order (NIEO), 167
New Zealand, 4, 19, 21, 30, 32, 41, 42, 44, 46, 52, 57, 65-6, 135, 144, 177
nickel, 71
nitrous oxide, 122
non-governmental organizations (NGOs), 127, 149
North Atlantic Treaty Organization (NATO), 140
North Sea Oil, 50
Norvegia, 88
Norway, 4, 19, 21, 30, 41, 46, 52, 57, 88
nuclear testing, 63, 97, 98, 100, 101, 102, 103, 104, 117, 140; ban on, 134
nuclear waste, dumping of, 79
nuclear-weapon free zones, 7, 21, 24, 89, 149
nuclear weapons, developed by South Africa, 98

O'Neill, Michael, 130
occupation of territory, 13
oil: costs of mining, 55; deposits of, 48, 51, 55, 66, 75; industry, 1, 24; prices of, 49, 114, 139; spills of, 2, 51, 53, 60-1, 62, 63, 136, 108-10
oil-drilling, 71, 114, 147, 162; technology, 50
Operation Gangotri, 73-5
Organization of American States (OAS), 131
Organization of Petroleum Exporting Countries (OPEC), 49
outer space, 14
over-fishing, 3, 36
Oxfam, 127
ozone layer, gap in, 5, 6, 7, 67, 95, 118-21, 123, 141, 144

Palme, Olof, 112
Palmer, Geoffrey, 65, 66
Papua New Guinea, 52
Pardo, Arvid, 69-71, 77, 160, 162, 166
Parfit, Michael, 180
Partial Test Ban Treaty, 140
peace movement, 140
Permanent Court of International Justice, 12
Peru, 23, 52, 73, 84
Pinto, Christopher, 72
piracy, 11
Poland, 22, 41, 42, 44, 52
Polar Circle, 74
Polar Year, 17
pollution, 5, 154; control of, 22
polymetallic sulphides, 167
prescription of territory, 13
Project Oasis, Australia, 154
public heritage, concept of, 169-80
public trust, concept of, 171

Rainbow Warrior, 102, 103
Raymond, Robert, 151
Reagan, Ronald, 140
Regulatory Committee, 56, 57
Renouf, Alan, 90
res communis, 163
res nullius, 162
research, 89
Resolutions of United Nations: 40/156B, 81; 40/156C, 81; 41/88A, 82; 41/88B, 82; 41/88C, 82; 42/46A, 82; 42/46B, 82; 43/83A, 83; 43/83B, 83; 44/124A, 84; 44/124B, 85
resources, ownership of, 91
Rice, Nan, 99
Rohl, Kenneth, 60
Romania, 46, 52
Rowland, Sherwood, 119
royalties on minerals, 50

satellite technology and national security, 175
Schneider, Stephen, 122
scientific research, 2, 4, 6, 30, 34, 47, 48, 50, 73, 85, 91, 97, 101, 107, 172, 174, 178
Scientific Committee for the Conservation of Antarctic Marine Living Resources, 40, 41, 42, 43
Scientific Committee on Antarctic Research (SCAR), 18, 32
Scientific, Technical and Environmental Advisory Committee, 56
Scullin, J.H., 87
sea, control of, 161
sea-bed, 14, 21; control of, 163; exploitation of, 3, 160; internationalization of, 69, 166
sea-levels, rising, 7

seal industry, 16, 32-4, 37, 47
seas, freedom of, 161
Selden, John, 161
skin cancer, 119
South Africa, 19, 22, 23, 41, 52, 57, 81, 83, 85, 86, 95, 96-100, 107, 173, 178
South Georgia, 41, 42
South Korea, 23, 31, 41, 57
Southern Raider, 103-4, 106
Spain, 41, 52
Sri Lanka, 72, 73, 86
Stockholm Conference, 138, 169, 170
strategic trust territory, Antarctica as, 91
Sweden, 52, 57
Switzerland, 18, 174

Taiwan, 31, 162, 163
Tasmania, 39, 63, 171
Taylor, Thomas N., 47
terra communis, 163, 165, 172
terra nullius, 162, 164-5
territorial limits, 13, 90, 161
territory: acquisition of, 13-15; control of, 13
Thatcher, Margaret, 6, 66-7, 75, 123
Thomson, Michael, 3
Torrey Canyon, 51, 138
tourism in Antarctica, 22, 150-2, 171, 172, 174, 178
tourist treaty, 150-2
trade unions, 129
transnational corporations, 70
treaties: bilateral, 11; multilateral, 11
treaty law, 11
Treaty on the Global Climate, 83
Triggs, Dr, 164

Union of Soviet Socialist Republics (USSR), 1, 7, 16, 18, 19, 21, 22, 29, 30, 32-3, 35, 37, 41, 42, 44, 47, 52, 53, 57, 71, 75, 77, 80, 81, 89, 91, 92, 98, 107, 122, 140, 144, 164, 175, 178, 180
United Kingdom (UK), 4, 16, 17, 19, 21, 22, 30, 33, 41, 42, 44, 46, 50, 52, 57, 66-7, 75, 88, 91, 92, 99, 102, 119, 121, 140, 161, 171, 177
United Nations (UN), 11, 22, 69-73, 91, 113, 115, 129, 144, 150, 170, 173, 174, 177; Charter, 10, 14, 91, 92, 96, 130, 174; Conference on the Human Environment, 49, 138, 139, 179; Economic and Social Council (ECOSOC), 130-1; Environment Programme (UNEP), 113, 120, 138; General Assembly, 1, 10, 75, 78-87, 95, 115, 131, 137, 139, 147, 173; International Law Commission, 70; Law of the Sea Convention, 12; Security Council, responsible for demilitarization, 91
United States of America (USA), 1, 7, 16, 18, 19, 20, 21, 22, 30, 41, 42, 46, 49, 52, 53, 57, 65, 66-7, 71, 75, 76, 77, 80, 89, 91, 92, 98, 102, 107, 108, 119, 128, 129, 130, 140, 141, 144, 151, 154, 164, 171, 175, 177, 178, 180
United Tasmanian Group, 63
uranium deposits, 48, 97
Uruguay, 52, 57, 73, 81, 110

Vallentine, Jo, 103
van Allen, Dr, 17
van Bynkershoek, Cornelius, 13
von Humboldt, Alexander, 17
Vostock station, 122

Wallace, Gilbert, 156
Warrick, Dr, 141
waste disposal, in Antarctica, 95, 108-10, 136
weather-reporting stations, 154
whales, 51
whaling industry, 4, 16, 28-32, 37, 47, 88, 162
Whitlam, Gough, 89
Wilenski, Peter, 84
Women's International League for Peace and Freedom, 142
Woolcott, Richard, 81, 83-4, 97
World Association for World Federation, 137
World Commission on Environment and Development (WCED), 115
World Conservation Strategy, 113-15
World Park, Antarctica as, 65, 85, 134-5, 144, 147, 169, 170, 179
World Vision, 127
World Wide Fund for Nature, 136
World Wildlife Fund, 113

Zaire, 64
Zambia, 64

The Antarctic and Southern Ocean Coalition

The Antarctic and Southern Ocean Coalition (ASOC) is a non-profit coalition comprised of more than 200 non-governmental environmental and conservation organisations (NGOs) from over 40 countries. Three of its member organisations act as 'secretariats' (The Antarctica Project in Washington DC, ASOC-NZ in Wellington, New Zealand, and Greenpeace Australia in Sydney).

ASOC has had an active interest in Antarctic issues, particularly concerning the impacts of human activities on the Antarctic environment, since 1977. The coalition was founded in that year, at the beginning of the negotiations for the Convention for the Conservation of Antarctic Marine Living Resources, for the following purposes:

- the distribution of information about Antarctic environmental and conservation issues, government policies and scientific issues to interested NGOs
- monitoring of meetings of the Antarctic Treaty, Convention on the Conservation of Marine Living Resources (CCAMLR), Scientific Committee on Antarctic Research, and other bodies that are involved with Antarctic environmental and conservation issues, and to report on those meetings to the world community
- preparation of proposals for improving pertinent international conventions and their implementation, including habitat sanctuaries, improved environmental impact assessment procedures, new and environmental protection institutions
- promotion of a dialogue among governments, scientists and official observer organisations on (1) the possibility of managing Antarctic fisheries as a large-scale scientific experiment, (2) the desirability of preserving the Antarctic region as a science preserve and wildlife sanctuary and (3) the use of Antarctica's unique demilitarised status and inspections provision as a model for arms control efforts in other parts of the world
- publishing of the international newspaper *ECO* whenever governments meet to discuss or negotiate Antarctic issues, and to distribute *ECO* to interested diplomats, scientists, academics and conservation leaders

- encouragement of more governments to take an active interest in Antarctica's future
- support of scientific research in the Antarctic, on a co-operative, international basis
- encouragement of international inspections to ensure compliance with agreed rules and measures, including establishment of an Infractions Committee for The Antarctic Treaty System.

The aims and objectives of ASOC, which are based on the principle that Antarctica's near-pristine environment and its current wilderness, aesthetic and scientific values should be permanently protected for the use and benefit of all humankind, are as follows:

- to work for the implementation of a conservation strategy for Antarctica, including the creation of the equivalent of an Antarctic Environmental Protection Agency
- to endorse the 'ecosystem as a whole' (as in CCAMLR) approach to conservation in Antarctica
- to support protection of the Antarctic as a science preserve and wildlife sanctuary, wilderness area and place of special aesthetic, spiritual and scenic values
- to oppose oil and minerals development in the region
- to maintain and build on the unique arms-free situation in Antarctica towards arms control on a larger scale, and
- to support new forms of specially protected areas, including whale habitat sanctuaries and 'wilderness' reserves.